The Gospel of Climate Skepticism

The Gospel of Climate Skepticism

Why Evangelical Christians Oppose Action on Climate Change

Robin Globus Veldman

UNIVERSITY OF CALIFORNIA PRESS

University of California Press
Oakland, California

Library of Congress Cataloging-in-Publication Data

Names: Veldman, Robin Globus, author.
Title: The gospel of climate skepticism : why evangelical
    Christians oppose action on climate change / Robin
    Globus Veldman.
Description: Oakland, California : University of
    California Press, [2019] | Includes bibliographical
    references and index.
Identifiers: LCCN 2019020049 (print) |
    ISBN 9780520303669 (cloth : alk. paper) |
    ISBN 9780520303676 (pbk. : alk. paper)
Subjects: LCSH: Climatic changes—Effect of human
    beings on—Religious aspects—Christianity. |
    Anti-environmentalism—Religious aspects—
    Christianity. | Christian conservatism—United
    States—History—21st century.
Classification: LCC GF80 .V45 2019 (print) |
    LCC GF80 (ebook) | DDC 363.738/74—dc23
LC record available at https://lccn.loc.gov/2019020049
LC ebook record available at https://lccn.loc.
    gov/2019980084

Manufactured in the United States of America

28 27 26 25 24 23 22 21 20 19
10 9 8 7 6 5 4 3 2 1

To my family, with love

Contents

Acknowledgments

About midway through this project, one of my daughters was diagnosed with a terminal health condition. Through an extraordinary feat of intelligence and perseverance, my husband saved her life. This book is dedicated to him, with my eternal gratitude, and to both my daughters, with so much love.

Given the interdisciplinarity of this work, I find myself especially indebted to the work of other scholars, especially in the fields of history, political science, and sociology. Among scholars of evangelical environmentalism, I am particularly indebted to the work of Laurel Kearns, whose insightful analyses of evangelical environmentalism are an incomparable resource; Katharine Wilkinson, whose book *Between God and Green* went with me nearly everywhere; and Lydia Bean and Steve Teles, whose report "Spreading the Gospel of Climate Change" crystallized so much for me about the backlash against climate change activism within the evangelical world. I have found myself equally indebted to and absorbed by the work of historians, most especially Daniel K. Williams, Matthew Avery Sutton, and George Marsden. Within environmental studies, Kari Norgaard's book *Living in Denial* vastly expanded my horizons when it came to understanding how the social world affects our conceptualization of climate change. It was an inspiration for everything I have subsequently written. From the world of sociology, Christian Smith and his colleagues' book *American Evangelicals: Embattled and Thriving* was equally transformative, as the following pages make clear.

The process of book writing has immeasurably deepened my appreciation for these insightful and meticulous works, as well as the many others I have been fortunate to be able to draw upon.

I wish to thank the scholars who corresponded with me personally. I thank Jean Blocker, Matthew Sutton, Robert Booth Fowler, Mark Silk, David Hackett, and Jacob Mauslein for answering my questions about aspects of their research or expertise that intersected with my own. Mark Ward Sr. generously shared his immense knowledge of the evangelical mass media with me. Theo Calderara offered insightful comments early in the book's gestation.

I was also fortunate that a number of scholars read and critiqued various chapters. Mark Peterson waded through a much longer version of chapter 1 with good spirit. Jim Proctor pushed me to sharpen the argument in chapter 2. Michael McVicar provided eloquent and thought-provoking commentary on chapter 4. Dara Wald bravely tackled a fifty-page version of chapter 7. Laurel Kearns sharpened the argument in chapter 8 immeasurably. Chapters 1 through 3 are improved and, I hope, less repetitive thanks to Sarah Pike's editorial guidance. Rachel Globus, my literal better half in this regard, is responsible for any liveliness readers may detect in the opening pages of the introduction and for the epilogue's coherence. Lucas Johnston provided keen comments on aspects of how my work related to the wider field of religion and nature, as well as comments on the epilogue. Leah Sarat and Bernard Zaleha have been invaluable sounding boards at various stages of the research process. Darren Sherkat, Gordon Globus, and two anonymous reviewers shared detailed and immensely useful comments on the entire manuscript. To all of you, thank you. I hope someday I can return the favor.

Throughout my time in academia, Bron Taylor has been a champion, guide, mentor, and friend. I am so glad you persuaded me to study at the University of Florida. Thank you for the many ways you have helped me since then. In addition to reading my work in dissertation form, Katrina Schwartz encouraged me to get over my fear of book writing. I thank Ken Wald, David Hackett, and Charles Wood for their mentorship and invaluable advice at the dissertation stage. To my fellow graduate students at UF, thank you for your energy, insight, and support.

As for research assistance, I was incredibly fortunate to have Stephanie Koontz at my side as I conducted field research in Georgia. Natalie Whitis provided indispensable and astute assistance with media analysis when I was at Iowa State. I also thank Keiame Lee Chong and Manual Grajales for assistance once I moved to Texas A&M.

I could not have completed this book without the support of my wonderful colleagues at Iowa State University. I especially thank Heimir Geirsson, Hector Avalos, Rose Caraway, Anne Clifford, Christopher Chase, Travis Chilcott, Kate Padgett Walsh, Cullen Padgett Walsh, and Janet Krengel for providing a stimulating and collegial academic home.

Two fellowships were essential to the completion of this project. The Tedder Family Doctoral Fellowship at the University of Florida provided support at the dissertation stage. A research grant from the Center for Excellence in the Arts and Humanities at Iowa State University also provided me with time to focus on turning the dissertation into a book manuscript.

In addition to family members already mentioned, I thank my mother, Maria Globus, for encouragement throughout the process, and James and Lorraine Veldman for countless hours of babysitting and for sharing your home with us so many times over the years. And Joseph, thanks for the many hours of conversation, debate, and editing. If you're not careful, I'll rope you into helping with the next book, too.

I thank Eric Schmidt, my editor at UC Press, for believing in this project and helping guide it to completion.

Finally, I am deeply grateful to the people I met in Georgia who generously shared their time and world with me, and to Jim Ball and Jonathan Merritt for permission to reprint "Climate Change: An Evangelical Call to Action" and "A Southern Baptist Declaration on the Environment and Climate Change."

Chapter 6 is a reprint of "What Is the Meaning of Greening? Cultural Analysis of a Southern Baptist Environmental Text," by Robin Globus Veldman. It was originally published in the *Journal of Contemporary Religion* issue 31, no. 2 (May 2016), and is reprinted here by permission of Taylor & Francis Ltd (http://www.tandfonline.com).

Abbreviations

Introduction

On March 7, 2012, the Republican U.S. senator from Oklahoma, James Inhofe, appeared on a Christian radio program called *Crosstalk*. One might expect that on a program whose goal was "to advance the gospel thru [sic] every legitimate means," Inhofe would have been invited to speak about his faith.[1] A member of the Presbyterian Church (USA) denomination, Inhofe would have had plenty to talk about—his four-year boycott of the Christmas parade in Tulsa, Oklahoma, after its name was changed to the more inclusive "Parade of Lights," for example, or his frequent trips to Africa that were inspired by "the political philosophy of Jesus."[2] But instead, Inhofe was there to promote a book he had recently published about climate change. The book's title, *The Greatest Hoax: How the Global Warming Conspiracy Threatens Your Future* (2012), riffed on the famous line, delivered on the Senate floor in 2003, that had established him as the Senate's best-known climate contrarian: "With all of the hysteria, all of the fear, all of the phony science, could it be that man-made global warming is the greatest hoax ever perpetrated on the American people?" he had queried. "It sure sounds like it."[3]

Inhofe had presented a number of arguments against climate change over the years, but he made a surprising addition in his 2012 radio conversation. Those who doubt that catastrophic climate change is real or that it is caused by human activities (positions I call *climate skepticism* in this book)[4] typically claim that climate models are inaccurate, that changes to the climate are part of natural cycles, that carbon dioxide is good because plants need it, that the supposed scientific consensus about climate change is false, or that global warming will be beneficial. But Inhofe's message that day pointed directly to faith. To understand what was happening with the climate, he argued, listeners should turn to the Bible, where God promises: "As long as the earth remains, there will be planting and harvest, cold and heat, summer and winter" (Gen. 8:22).[5] For Inhofe, the passage was directly applicable to climate change. "My point is," he told listeners, "God's still up there. The arrogance of people to think that we, human beings, would be able to change what He is doing in the climate is to me outrageous."[6] Why was Inhofe talking about climate change—an issue that would appear to have little to do with religion—on a Christian radio program? Did he truly believe that the Bible was a good source for information about the future trajectory of global weather patterns?

It is tempting to dismiss Inhofe's apparently faith-based climate skepticism as an anomaly—pandering to his base, perhaps, or an idiosyncratic theology crafted to buttress his anti-environmental policy positions. But Inhofe is far from alone in holding such views. Polls have consistently found that white evangelical Christians are the most skeptical major religious group in the country when it comes to climate change. According to a survey conducted by the Pew Research Center in 2014, for example, just 28 percent of white evangelicals accepted that the climate was changing as a result of human activities, compared to 41 percent of white mainline Protestants, 56 percent of black Protestants, and 77 percent of Hispanic Catholics.[7] Nor was this survey anomalous, for in the 2000s white evangelical Protestants' levels of acceptance of climate change have consistently been lower than those of other religious groups in America.[8] Not only is being evangelical associated with lower levels of concern, but among evangelicals concern about climate change and religious commitment are inversely related: the most religious evangelicals are the least likely to be concerned about climate change.[9]

This skepticism cannot be completely reduced to other factors. Social scientific studies have consistently shown that even after controlling for factors such as political orientation, education, or media exposure, there is still a statistically significant association between evangelicalism and climate skepticism.[10] In fact, evangelicalism is one of just a few religious traditions (and in some studies the *only* one) to maintain a statistically significant association with climate skepticism after controls are added.[11] Nor is the effect minuscule. Republicans are in general more likely than Democrats to be climate skeptics, but a self-rated "strong" Republican of high religiosity is more than twice as likely as a "strong" Republican of low religiosity to deny the human origins of climate change.[12]

Inhofe, in other words, had plenty of company in his views, company whose views seem to be similarly rooted in faith. For those of us who hope that actions will be taken to address climate change, this raises several questions: Why do white evangelicals in America tend to be so skeptical about climate change? What does it have to do with their religious beliefs? What does this portend for U.S. participation in efforts to halt or slow the effects of climate change? Is climate skepticism destined to become as entrenched in evangelical circles as disbelief in evolution?[13]

This book offers answers to these questions. I explain how climate skepticism has come to feel like the "natural," "normal," or even "Christian" position to a subset of theologically and politically conservative evangelicals from historically white denominations called "traditionalists"—even though, as I argue, the link between faith and climate skepticism is not inevitable. While observing the ways in which climate skepticism has become tied to the traditionalist evangelical identity and reinforced by social dynamics, I also show how this seemingly spontaneous attitude has, in fact, been heavily promoted by leaders in the politicized arm of the evangelical tradition, often referred to as the Christian Right.[14] In other words, in connecting faith to climate change skepticism, Inhofe was not the exception, but something more like the rule.

HOW THIS BOOK BEGAN

I did not start out intending to write a book about the environmental attitudes of traditionalist evangelicals—that is, evangelicals who generally support the movement within evangelicalism known as the Christian Right—but, rather, gradually realized their significance as my research progressed. My initial focus was on the broader evangelical tradition, which should be carefully distinguished from the Christian Right. Evangelicalism can be defined as Protestants who affirm a belief that lives need to be transformed through a "born-again" experience and a lifelong process of following Jesus; who believe they must express and demonstrate the gospel in missionary and social reform efforts; who have a high regard for and obedience to the Bible as the ultimate authority; and who stress the sacrifice of Jesus Christ on the cross as making possible the redemption of humanity.[15]

The term *Christian Right* refers not to this broader group, but to a politically conservative subset of evangelical leaders allied closely with the Republican Party. Although it has deeper historical roots, this alliance was formalized in 1980 during the candidacy of Ronald Reagan when the Republican Party altered its platform to endorse conservative evangelical concerns like restoring prayer in public schools and prohibiting abortion.[16] Not all evangelicals support the Christian Right, however—some because they do not share its goals for society (traditional family values and free-market economics), and others because ethnic and racial ties have led them in different directions politically (e.g., many black and Latino Protestants).[17]

While generally socially and politically conservative, the evangelical tradition in America is larger and more diverse than the Christian Right, encompassing some ninety million to one hundred million Americans.[18] It is difficult to pin down exactly how many of these believers support the Christian Right, because of shifts over time and different measurement strategies. But it is very clear that not all evangelicals support the Christian Right. A 2010 survey found, for example, that just 29 percent of white evangelicals did so.[19]

While the Christian Right tends to shape the public perception of evangelicals as right-wing Bible-thumpers, social scientific studies reveal more diversity in their views. Among different groups within the evangelical tradition (e.g., Pentecostals and Baptists), for example, the percentage of those who believe that abortion should be legal in all or most cases ranges from nearly a third among nondenominational evangelicals to fully half among Presbyterian evangelicals.[20] Evangelicals' political affiliations are surprisingly diverse as well. Within the Churches of Christ (an evangelical denomination), a 2014 survey found that just half self-identified as Republican or leaning Republican, while 39 percent identified as Democratic or leaning Democratic and 11 percent as independent or leaning independent. Even among Southern Baptists, a historically conservative denomination that has had close ties to the Republican Party, a full 26 percent identified as Democratic or leaning Democratic.[21] On almost every issue, American evangelicals' views are less monolithic than is commonly supposed.[22]

When I moved to Georgia in the summer of 2011 to begin field research, I was not thinking narrowly about evangelicals who supported the Christian Right, but about the broader evangelical tradition. My goals were simple. On one hand, aware of studies that had shown evangelicals to be less concerned about the environment than other Americans, I hoped to better understand the sources of their apparent indifference. On the other hand, wary of the caricatured ways in which evangelicals are often portrayed by environmentalists and the mainstream news media, I hoped to provide a more nuanced portrait of their environmental views.

These questions were especially timely when I began my project because American evangelicalism appeared to be in the midst of a "greening" trend. A major evangelical environmental organization, the Evangelical Environmental Network (EEN), had been founded in 1993.[23] In the early 2000s, it decided to tackle climate change, with the goal of raising concern about this issue within the evangelical mainstream.[24] A number of signs indicated that its prospects for success were good. In 2004, the National Association of Evangelicals (NAE) had announced *creation care*—evangelicals' preferred term for environmental concern—as one of its seven priorities for public engagement. To observers' surprise, key leaders in the Christian Right endorsed the statement. The same year, *Christianity Today,* a leading evangelical publication, came out in support of a global warming bill—one that, observers noted, the Bush administration opposed.[25]

Then, in February 2006, the EEN and a number of other prominent evangelical leaders, working under the auspices of the Evangelical Climate Initiative (ECI), issued a statement called "Climate Change: An Evangelical Call to Action," which boldly asserted that evangelicals "must engage this issue without any further lingering" (see appendix C).[26] Signed by eighty-six influential evangelical leaders and announced with full-page ads in the *New York Times* and *Christianity Today,* the statement made national headlines.[27] The idea that the evangelical tradition was rapidly greening gained further credibility when Southern Baptists, a denomination with a strong reputation for conservatism, released what appeared to be a parallel statement encouraging action two years later.[28] Critically, the two statements seemed to indicate that evangelicals were breaking ranks with Republicans on the issue of the environment, perhaps clearing the path to legislative action. Thus, I entered the field at a moment of heightened curiosity about evangelicals' attitudes toward the environment.

Given that many religious traditions have produced official statements regarding climate change and other environmental issues, but few have engaged in any further action or activism, I envisioned using empirical research to judiciously evaluate claims about the greening of evangelicalism.[29] While, as an environmentalist myself, I was heartened by the apparent surge in environmental concern, I noticed that most observers (whether scholarly or journalistic) had focused on the actions of evangelical *leaders*. To understand the greening movement's prospects, it seemed critical to learn whether the activity at the top was resonating at the grassroots level as well. Hence, I determined to focus on the views of the evangelical laity.

As for my approach, I did not want to simply pass judgment about how environmentally conscious evangelical laypeople were. Rather, following a general practice in religious studies—that of attempting to see the world through the insider's eyes—I wanted to make sense of evangelicals' attitudes toward the environment on *their* terms.[30] In doing so, I hoped to sidestep the practice, common in social scientific studies of environmental attitudes, of adopting what was essentially the environmental movement's own definition of environmental concern as the yardstick. The problem with the latter approach is that the environmental movement has a specific history, which shapes its vision of appropriate environmental concern. That history is particularly problematic when it comes to evangelicals because environmentalists have long suspected Christianity of encouraging anti-environmental sentiments. This suspicion seems to have shaped some definitions of environmental concern in ways that make it difficult for evangelicals to be counted as environmentally concerned.

One example of this can be seen in the New Ecological Paradigm scale, which is one of the most commonly used tools for measuring pro-environmental orientation. One of the ways the New Ecological Paradigm scale measures environmental concern is by asking respondents whether they think humans were meant to rule over the rest of nature.[31] Agreement with this idea is coded as anti-environmental in the scale, yet the idea itself comes straight from the Bible. In Genesis 1:28, God tells mankind to "fill the earth and subdue it. Rule over the fish in the sea and the birds in the sky and over every living creature that moves on the ground."[32] Since one of the defining features of the evangelical tradition is according high authority to scripture, the item verges on coding evangelicalism itself as anti-environmental. At best this particular item leaves little room for evangelicals (especially those outside of erudite circles) to be considered environmentally concerned while still expressing traditional attitudes toward the Bible.

Finally, I was also interested in the social aspects of environmental concern, and in how churches as social environments might play a role in enhancing or inhibiting concern. Political scientists had established in the 1980s that even without including overt political content in sermons, churches managed to transmit political attitudes to members, resulting in a tendency for members' views to converge over time.[33] I hoped to understand the extent to which the same dynamic might be at play when it came to environmental attitudes. It may seem obvious that evangelicals would absorb environmental attitudes from the people sitting next to them in the pews, but for historical reasons, most of the social scientific research on Christians' (and, by extension, evangelicals') environmental views has focused on the individual-level impacts of Christian theology. Regrettably, in this line of research, the social world has remained almost entirely invisible.[34]

While my approach was guided by these three concerns—to be appropriately critical of claims about the "greening" of evangelicalism, to convey evangelicals' views in their own words, and to be attentive to social dynamics—my central question was more specific: To what extent, if any, were evangelicals' attitudes about the environment shaped by their beliefs about the end times? Although it is perhaps not widely known, a number of prominent environmentalists, including Al Gore and E. O. Wilson, have asserted that the belief that Jesus could return to earth at any moment is a critical factor undermining evangelicals' concern about the environment. As the journalist Bill Moyers asked in a speech that circulated widely in the mid-2000s, "Why care about the earth when the droughts, floods, famine, and pestilence brought by ecological collapse are signs of the apocalypse foretold in the Bible? Why care about global climate change when you and yours will be rescued in the Rapture?"[35] Those raising such concerns often point to polls revealing the popularity of such beliefs: a 2010 poll conducted by the Pew Research Center, for example, found that 58 percent of white evangelical Christians believed that Jesus would return by 2050.[36] From this, many environmentalists draw the lesson that, in the words of the well-known environmental educator David Orr, "belief in the imminence of the end times tends to make evangelicals careless stewards of our forests, soils, wildlife, air, water, seas and climate."[37] Despite such allegations, at the time when I was conceiving of my project, just a handful of studies had sought to examine the connection between end-time beliefs and environmental attitudes empirically, and none of them had involved field research. Hence, I went to the field hoping to shed light on whether (and, if so, to what extent) it was true that end-time beliefs caused environmental apathy, a theory that I refer to as the *end-time apathy hypothesis*.

The end-time apathy hypothesis is, to be sure, an idiosyncratic starting point for a project on evangelicals' environmental attitudes, especially when factors like political views are more commonly cited as discouraging concern. But my goal was never only to investigate end-time beliefs, nor did I believe they were necessarily the most powerful factor diminishing concern. The end-time apathy hypothesis was simply a starting point, an intriguing and little-explored piece of the puzzle that I was eager to better understand—alongside any additional factors I might uncover.

During a year conducting qualitative field research in Georgia (explained further below and in appendix A), I came to see why so many social scientific studies have found evangelical or fundamentalist Christians to be relatively unconcerned about the environment.[38] My informants struggled, in many cases, to even name existing environmental problems, and as we talked about environmental problems, many of them went further, expressing skepticism and hostility toward the environmental movement itself. The issue of climate change was no exception; indeed, it almost universally provoked condemnation. I once had a woman refuse to talk to me after I tried to clarify that global warming was different from the hole in the ozone layer (a misstep that I never repeated). As a colleague of mine said of a similar research project (also conducted in Georgia), asking about climate change was "perhaps the most difficult research experience of my life. . . . I hadn't even gotten through a couple of sentences of my intro and [my informants] were all over me." Perhaps because many of my conversations were conducted in church, my informants tended to be extremely polite. But they did not hesitate to ridicule, denounce, and revile those who worried about human activities permanently altering the planet's climate.

Nevertheless, it became clear to me that the end-time apathy hypothesis was not the best way of conceptualizing the relationship between my informants' faith and their environmental attitudes. A small number of the evangelicals I met were so convinced the end is near that they cared little about environmental decline, viewing it as one of many indicators that history was drawing to a close. But a larger, more politically engaged contingent was convinced that the climate was not changing at all—or, if it was changing, that humans were not the cause and/or that the changes would not be catastrophic. For these individuals, climate change was not evidence that the end times were beginning, but instead a hoax—a competing eschatology concocted by secularists who sought to scare people into turning to government instead of God.[39] I concluded that this *sense of embattlement with secular culture* explained why so many of my informants rejected climate change on religious grounds. They were not climate skeptics only for this reason, of course. Rather, I found that political and social conservatism, attitudes toward collective action, community norms, and media consumption habits all encouraged climate skepticism. But it was the "embattled" mentality that seemed to explain what climate change skepticism had to do with their faith, solving a significant puzzle in the literature on evangelical climate change skepticism.

Evangelicals' embattled mentality has deep roots in the tradition's history. According to the historian George Marsden, in the nineteenth century, evangelical Protestants held influential positions in American society and consequently considered themselves the nation's moral guardians.[40] Since the turn of the twentieth century, however, this privileged position has been slipping away. Upset by this and other changes, in the latter half of the twentieth century a group of conservative evangelical leaders began to rally evangelicals in defense of traditional values, in part by arguing that Christianity was under attack by secular humanists. "It is no overstatement to declare that most of today's evils can be traced to secular humanism," wrote the prominent evangelical author and leader Tim LaHaye in 2000. Indeed, humanists had "already taken over our government, the United Nations, education, television, and most of the other centers of life."[41]

By the end of the twentieth century, leaders like LaHaye had managed to make the battle with secular culture a major issue within evangelicalism and, in doing so, to mobilize a mass movement that seemed to be energized, rather than demoralized, by pluralism and modernity. Where scholars had once expected the evangelical tradition to fade away, sociological research conducted in the late 1990s established that in fact, the tradition was flourishing—and not despite modernity, but *because* of it. Contrary to expectations, doing battle with secularists and competing religious groups in American society had both focused and energized evangelicals, leading the sociologist Christian Smith and his colleagues to pronounce the tradition "embattled and thriving."[42]

This sense of embattlement with secular culture, it turned out, had major consequences for attitudes toward climate change. Many of my informants rejected the notion that humans could be changing earth's climate because they viewed it as an attempt to deny God's omnipotence. As I heard repeatedly, "God is in control"—of the climate and of everything else. Echoing Inhofe, they argued that it was arrogant to assert that humans, rather than God, controlled the earth's destiny. In this context, rejecting climate change became a means of holding fast to one's faith in a hostile, rapidly secularizing world. "I know how God says the earth is going to end," a middle-aged evangelical woman confidently explained to me, in reference to biblical end-time teachings. "I don't sit around fearing that the polar ice caps are melting."

The sense of embattlement was powerful not only because it could neutralize concerns about climate change by explaining it in terms of something else—a secularist plot—but because it was sustained by social dynamics. For many evangelicals, being theologically and politically conservative is a central aspect of their identity. This made embracing a politically liberal issue like climate change socially risky. If someone could embrace a politically liberal issue, did that mean he or she was also willing to embrace liberal theology? In such a polarized social context, to express concern about the environment or climate change was to put one's reputation on the line. Even powerful leaders were subject to such forces. "I have been called names that I have not been called in my entire life," complained then Southern Baptist Convention (SBC) president Frank Page in 2008, after he joined an effort to raise awareness about climate change in his denomination. "We have been accused of being a part of a left wing, liberal agenda," he elaborated. "There has been much misunderstanding, much anger, and much speaking of unkind and demeaning words."[43]

As a researcher, I wondered whether what I was hearing was a local phenomenon or part of a broader evangelical discourse. As other studies on the topic were published, it became increasingly evident that the latter was the case. As chapter 6 describes, qualitative studies conducted with evangelicals in other parts of the country and national-level surveys provide evidence that the kinds of comments I heard in Georgia could be heard in any number of evangelical churches around the country. To be sure, only a minority of evangelicals holds these views. Yet they are worth attending to, because this minority is vocal and politically engaged.

To return to the story of my research, while increasingly convinced that what I had encountered in Georgia was part of a broader national pattern, I still did not fully understand where these attitudes came from. In seeking to better fathom the embattled mentality's scope, I stumbled across the answer. I had assembled a list of what evangelical leaders said about climate change, testing the hypothesis (simplistic as it now seems) that those leaders who endorsed action on climate change did not hold the embattled mentality, while leaders who opposed it did. As the list grew longer, I realized not only that all the skeptical voices were associated with the Christian Right, but that these individuals had been actively engaged in promoting those views to the laity. Moreover, the language these leaders were using strongly echoed what I had heard while in the field.

After conducting a more systematic analysis, it began to make far more sense to me that so many evangelicals around the country were dismissing climate change in religious terms, for this was exactly what leaders in the Christian Right had been urging them to do. Importantly, these leaders had a powerful tool at their disposal: the evangelical mass media. Composed of radio, television, and digital media, the evangelical mass media reach millions of evangelical Christians around the nation daily. Indeed, an estimated one in five Americans consumes these media daily, and 90 percent of evangelicals do so monthly.[44] Via such means, influential leaders like Jerry Falwell, Pat Robertson, Chuck Colson, Ken Ham, and many others have been able to promote climate skepticism widely.

Why, after decades of mostly ignoring environmental issues, did leaders in the Christian Right decide to get involved in the debate over climate change? Their allliance with Republicans is relevant, but the more proximate factor was the ECI. The Christian Right's precarious hold over the evangelical tradition in the mid-2000s, when the push for action on climate change emerged, coupled with their need to support coalition partners in the Republican Party who opposed action on climate change, forced them to directly involve themselves in the climate fight.[45] They worked to thwart the ECI's momentum, and, with the power of the evangelical mass media at their disposal, they largely succeeded.

To say this is not to imply that traditionalist evangelicals are climate skeptics only or even primarily for religious reasons.[46] Rather, I am arguing that the leaders in the Christian Right who promoted climate skepticism did two things. First, they supported and amplified the efforts of secular climate change skeptics, ensuring that their message reached evangelical audiences.[47] Second, many of their communications explicitly framed climate change as a religious issue, and in so doing helped transform climate skepticism and denial from a political opinion into an aspect of evangelical identity. This transformation had serious repercussions. Most critically, the mocking, disdainful, and suspicious tone that leaders and pundits in the Christian Right adopted when speaking about climate change sent a clear signal to evangelical audiences not only that skepticism was the more reasonable position, but more subtly that it was the more socially acceptable position. The result was that anyone considering expressing a different opinion knew that they risked being challenged or viewed with suspicion themselves. When climate advocates became enemies and climate advocacy became suspect, the greening of evangelicalism, which had been a growing chorus, fell to a whisper.

THIS BOOK'S CONTRIBUTION

While a number of valuable studies of evangelical environmentalism already exist, my argument improves our understanding of evangelicals' attitudes toward the environment and climate change in important ways. Previous research has attributed evangelicals' skepticism to four main factors. The first is *politics:* numerous scholars and observers have argued that evangelicals' skeptical attitudes toward climate change are rooted in, or may even reduce entirely to, their political conservatism.[48] Second, some scholars have argued that *anti-science attitudes* rooted in the creation-evolution debate contribute to skepticism.[49] Third, scholars have also contended that evangelicals' reluctance to support action on climate change derives from their preference for *individual over collective action.*[50]

Historically speaking, however, no factor has received more attention when it comes to explaining evangelicals' attitudes toward the environment than *theology.* Most famously, in 1967 the historian Lynn White Jr. accused Christianity (in general) of precipitating the environmental crisis by being "the most anthropocentric religion the world has seen."[51] For White, the creation story in Genesis was particularly to blame, for it depicted God creating the world "explicitly for man's benefit and rule."[52] Scholars have hypothesized that this reading of Genesis helps explain conservative Protestants' lack of environmental concern compared to other Christian groups, since conservative Protestants grant greater authority to scripture.[53] In addition to biblically rooted anthropocentrism, other proposed theological barriers to environmental concern for evangelicals include the fear that environmentalism will lead to paganism, pantheism, or other forms of earth-worship;[54] the argument that Christians should focus on salvation and saving souls, rather than pursuing a broader social agenda;[55] and the belief in Jesus's imminent return.[56]

What is missing from these explanations is an account of how evangelicals' environmental attitudes have been shaped by the movement's history, goals, and social practices.[57] Despite what partisans would like us to believe, the Bible's teachings on how humans should treat the natural world can and have been interpreted many ways. The question is how have they been interpreted by traditionalist evangelical Christians in early twenty-first-century America, and why did they adopt these specific interpretations over others? I argue that to understand this at the level of evangelical elites requires understanding *history*—particularly the origins of evangelicals' sense of marginalization and how the political strategy that leaders in the Christian Right adopted to combat it forced them to directly confront those advocating action on climate change in the mid- to late 2000s. At the level of laypeople, it requires understanding *identity*—particularly how and why rejecting environmentalism and concern about climate change came to be seen as a way of affirming one's identity as a Bible-believing Christian. Tying these two together is the evangelical mass media, which allowed leaders and pundits in the Christian Right to both cultivate a particular vision of evangelical identity as normative and convince evangelical audiences that certain environmental attitudes were the natural expression of this identity.

Unfortunately for those of us concerned about climate change, this underlines a key finding from the literature on risk perceptions. When it comes to climate change, risk perceptions are not guided simply by the magnitude of the possible catastrophe, but by its "congruence with individuals' cultural commitments."[58] Indeed, traditionalist evangelicals may constitute the paradigmatic example of this phenomenon.

The perspective offered here also answers calls to move beyond the Lynn White thesis that has dominated discussions of religions and the environment.[59] Against the tendency to view Christians' environmental attitudes as the product of ahistorical religious doctrines, I underline that it was not inevitable that so many evangelicals would come to regard climate skepticism as a matter of faith. This view was cultivated by specific individuals operating in unique historical and political circumstances in order to achieve specific ends. I hasten to add that this is not merely a cynical view of religion and religious actors, but one that invites a dynamic conception of how religions are related to environmental attitudes and behavior. Not by nature, but by convention. And what is conventional can be changed.

METHODOLOGY AND METHODS

My research trajectory has been winding and requires some explanation of the methodology underpinning it. The strategy I deploy is rooted in a tradition of qualitative research known as grounded theory, which was first formulated in the late 1960s as a way of giving inductive, qualitative research a high level of rigor, one comparable to that attained by deductive, hypothesis-testing studies. Grounded theorists do not formulate hypotheses and test them; they engage in open-ended but directed encounters that allow informants to express their own perspectives and views of the world.[60] These narratives, in turn, become the data that the researcher uses to formulate preliminary explanations, which then inform subsequent interviews and interpretations. Hence, rather than imposing hypotheses about the data at the outset, hypotheses are revised throughout the research process until they hold true for all of the evidence; through continuous revision, the researcher's understanding becomes deeper, richer, and better "grounded" in the data.[61] Among the many advantages of this approach was that it allowed me to start with something of a caricature as a research question, but without limiting my conclusions to that simplistic conception.

Typically, the researcher's questions or hypotheses dictate a particular set of methods, but since I revised and refined my hypotheses during the course of research, I found it necessary to employ different methods during each phase of research. In the first phase, I used ethnographic and qualitative methods to construct a nuanced portrait of traditionalist evangelicals' engagement with the environment and climate change in Georgia. I focused primarily on historically white evangelical denominations because this was the group that surveys and social scientific studies had identified as having consistently low levels of environmental concern (as a result, 94 percent of my focus group participants and all pastors but one were white). During this phase, I lived and conducted fieldwork for fourteen months in a town of about thirty thousand people in Georgia. To protect the privacy of those who were gracious enough to speak with me during my time there, I will refer to the town by a pseudonym, Mayfield. I spent about half of the time I lived in Mayfield arranging a series of nine focus groups at evangelical churches of various denominations in the area, including (for balance) some in nearby rural, suburban, and urban areas. The urban area, which I will call Dixon, had a population of more than fifty thousand. I selected focus groups as my primary research tool for the ethnographic phase because they are well suited to assessing people's perceptions and feelings about a particular topic, uncovering differences in perspectives between groups or categories of people, and revealing factors that influence opinions, behavior, and motivation.[62] All focus group participants also completed a background survey that collected basic demographic information as well as information about frequency of church attendance, Bible views, occupation, political party, and political ideology.

I began the process of setting up a focus group by interviewing pastors, who served as gatekeepers to the communities. In addition to interviewing pastors, at every church where I conducted a focus group I also attended at least one and often several services; when possible, I also attended Wednesday evening Bible studies. I also selected one church, a rural fundamentalist one that I will call Pinewood Baptist Church, for extended study. Over a five-month period, I attended Sunday services, some Wednesday evening Bible studies, and a Vacation Bible School at Pinewood Baptist. There I was lucky enough to meet a middle-aged high school educator I will call Jill. Having been raised in the Pacific Northwest, Jill moved to rural Georgia in her late teens, lived there for the succeeding twenty-five years, and befriended the pastor of Pinewood Baptist. As someone who had deep knowledge of the local culture but also retained an outsider's perspective, she was an articulate and perceptive observer of local cultural and religious norms. Fortunately for me, she was also in the midst of pursuing a graduate degree in environmental education, so she had an excellent grasp of both evangelical and environmental perspectives. Jill added greatly to my understanding of local attitudes.

To attain greater breadth, I also interviewed neighbors and others I met about their experiences living in the area, their understanding of the role of faith in the region, and their sense of how people around them regarded the environment and climate change. For balance, I also conducted participant observation and interviews with a number of environmentalists who worked or volunteered in the region. These interviews allowed me to talk my hypotheses over with people who were familiar with the local culture and, in doing so, validate and refine my conclusions. Hence, while the focus groups form the core of the analysis, my conclusions are drawn from a range of perspectives.

The second phase of research, on Southern Baptists, drew on both interviews and tools for textual analysis developed in the field of cultural studies (see chapter 6). The final phase of research, presented in chapter 7, entailed qualitative data analysis of digital archives maintained by leaders and pundits associated with the Christian Right. Appendix A describes each of these phases in greater detail.

If the primary method utilized in each phase differed, the plan of analysis did not: grounded theory was the methodological chain linking each step. Regarding the result, I am reminded of something the geographer Clarence Glacken wrote in the introduction to his magisterial work on ideas about nature in the Western tradition. The interdisciplinary nature of his line of inquiry led him, he reported, "into chilly . . . regions whose borders are patrolled by men who know every square foot of it."[63] My methodological trek has required a similar amount of trespassing. However, convinced as I am that the phenomenon of evangelical climate skepticism cannot be understood—or, at least, not understood well—without a deeper view of evangelical history and sociology, I think this trespassing was worth the risk. I hope readers will agree.

Finally, since I set out to examine the beliefs and values of a group of which I am not a part, some readers may appreciate an explanation of my motives and approach. In considering the work of other scholars in my position, I have found a particular affinity with the "critical empathy" approach of the religious studies scholar R. Marie Griffith. According to Griffith, critical empathy entails promoting an accurate, even sympathetic, understanding of one's informants—but not at the sacrifice of critical analysis. In her study of a charismatic evangelical women's group, Griffith writes that she adopted this approach in order to push back against the "flat, stereotypical terms" that outside observers typically applied to her informants.[64] Finding that environmentalists often tend to view evangelical Christians in the same terms, I have similarly labored to present my informants' views in detail, considering both the discrepancies and the too-often-ignored points of overlap between the environmental values of my informants and of environmentalists.

Like Griffith, however, I have also found it necessary to "[bring] some critical voices to light."[65] Griffith writes that she came to admire her informants' "refusal to give in to the easy cynicism of our time."[66] I did admire and even envy some aspects of the faith communities I visited, particularly the support they provided to those in their communities who were suffering. But I saw little to admire in my informants' hostile attitudes toward environmentalism and climate change. Indeed, my view is that my informants have been seriously misled about both, and one of the purposes of this book is to make clear who is responsible for such misapprehensions.

DEFINING "TRADITIONALIST" EVANGELICALS

This study is about a subset of evangelicals called traditionalists. Before I describe them, however, I want to clarify some of the other terminology used to describe theologically conservative Protestants. It is easy to become confused about who this group is because so many terms are in use, and because the usage is imprecise, especially in press accounts. Confusingly, terms like *evangelical Christian, Christian Right, Religious Right, conservative Christian, born-again Christian, conservative Protestant,* and *fundamentalist* are used interchangeably, when they properly refer to different groups.[67] That said, coming up with the appropriate terminology is not easy. In fact, how to best define and measure evangelicals is a point of significant debate among researchers and political observers alike, not only because it is important to be accurate but also because evangelicals' rise to political prominence in the late twentieth century has rendered claims about the size of their membership politically significant.[68]

Many sociologists and political scientists advocate using *conservative Protestant* as an umbrella term for the four major groups of theologically conservative Protestants in North America: evangelicals, fundamentalists, charismatics, and Pentecostals.[69] This book is about conservative Protestants, but as none of my primary sources or informants use the term, I have found it exceedingly awkward to use. Hence, I employ *evangelical* as the umbrella term, although it, too, is problematic, primarily because it has been used in so many different ways. The best way I've found to clarify what this term and the others it is often interchanged with mean is to turn to history. I will do so briefly, since many excellent, detailed histories are available.[70]

To *evangelize,* in its original sense, simply means to spread the good news of salvation. Although, by that definition, the term could theoretically apply to all Christians, by the eighteenth and nineteenth centuries *evangelical* had become the common name in both Britain and America for the revival movements of the period, which affected virtually all Protestant denominations. At this point, evangelicals enjoyed a great deal of societal prestige as the spiritual core of an essentially Christian nation. Evangelical Protestants were leaders in politics, education, and business, and Christian values largely set the agenda of the developing American culture. In public schools throughout the country, children were taught the Ten Commandments and read scripture as part of basic moral instruction.[71] However, this prestige was about to undergo a series of shocks and challenges that would alter evangelicalism's course profoundly over the next century.

Foremost among these challenges were new intellectual developments. First, in the late nineteenth century, divisions emerged over how to interpret the Bible. Those who were theologically liberal embraced a new method of analysis known as "higher criticism," which treated the Bible as literature, and in so doing raised questions about its authorship, historicity, and dating. Calling themselves "modernists," many theological liberals embraced the perspective afforded by higher criticism as a new stage in the evolution of the Christian tradition.

Charles Darwin's book *On the Origin of Species* (1859) further challenged the faith by questioning the biblical account of creation. While modernists sought to reinterpret scripture so that it would harmonize with modern science, theological conservatives came to reject "Darwinism," which they viewed as a hypothesis rather than scientific fact. So convinced were they of its pernicious effects on Christian faith and morality that in the early twentieth century they sought, famously and unsuccessfully, to prevent evolution from being taught in public schools.

With tension mounting, conservatives attempted to force liberals out of their denominations.[72] When that failed, conservatives launched another offensive, publishing twelve paperback volumes defending *The Fundamentals* of the Bible between 1910 and 1915. During this period some three million copies were distributed to pastors, theology professors and students, religious editors, and other religious leaders, enormously boosting the conservative perspective.[73] The books also supplied the name for the emerging movement of theological conservatives. In the words of the Baptist minister and editor Curtis Lee Laws, who first proposed it, the moniker *fundamentalist* would apply to those "who mean to do battle royal for the fundamentals" of the Christian faith.[74] It was not only a name, but a declaration of war. Fundamentalists of today, who trace their history to this early twentieth-century reaction against modernism, still emphasize a strict literal interpretation of the Bible (including its teachings about creation, the end times, and miracles) and institutional separation from liberal Protestants and Catholics, whom they view as apostates.[75]

Evangelicalism, in the contemporary sense of the word, grew out of frustration with the fundamentalist movement. In the first quarter of the twentieth century, fundamentalism gained a negative connotation, being associated (among other things) with racism, ignorance, and intolerance.[76] To counter that impression, and to provide a mechanism for cooperation across regional and denominational boundaries on issues of common interest, in 1942 a group of moderate (compared to fundamentalists, at least) conservative Protestants joined together to form the NAE.[77] Evangelicalism in its modern sense descends from that effort.

As for the two other Christian movements that belong under the conservative Protestant umbrella, *Pentecostalism* coalesced in 1901 under the leadership of Charles F. Parham and William J. Seymour and emphasized speaking in tongues as a key element of Christian life.[78] More than other evangelical movements of the time, Pentecostalism cut across racial lines—a heritage that was still evident in my field research in Georgia. *Charismatic renewal* (also called neo-Pentecostalism), which emphasizes speaking in tongues and miraculous healing, developed within conservative Protestantism during the 1960s and spread within Catholicism as well as within mainline and evangelical denominations.[79]

Traditionalist evangelicals, who are the focus of this study, are a subset of conservative Protestants from historically white denominations who have become politically engaged with the hope of restoring Christian values to their central place in American culture and law. They can be found in evangelical, fundamentalist, or Pentecostal denominations and within the charismatic movement. According to the political scientist John C. Green, who uses the term in his survey research, among evangelicals, traditionalists "[come] closest to the 'religious right' widely discussed in the media."[80] They tend to be Republican, to have a high view of the authority of the Bible, attend church regularly, believe that religious groups should be politically involved and affirm that their religious beliefs affect their political views, and (critically) desire to preserve traditional beliefs and practices in a changing world.[81] In my focus groups, 72 percent identified themselves as Republican (compared to 70 percent in Green's study); 85 percent agreed that the Bible "is the actual word of God and is to be taken literally, word for word"; and 96 percent attended church at least once a week. From their comments, as well, it was clear that they opposed the marginalization of traditional Christian belief and practice.

While not all white evangelicals are traditionalists, traditionalists are some of the loudest voices within the evangelical community. Indeed, the Christian Right's clout is due, in part, to its claim to speak for this sizable group.[82] In Green's nationally representative 2004 survey, traditionalists made up 12.6 percent of the American population, which constituted almost half of all evangelicals. They were the largest of the eighteen religious groups he measured as comprising America's religious landscape. In 2008 they appeared to have declined slightly, making up 10 percent of the general population and 40 percent of evangelicals.[83] By 2017, however, their numbers appeared to have grown (perhaps having been mobilized during the 2016 election). In 2017 the Baylor Religion Survey found that among white evangelical Protestants (measured by denominational affiliation), 72 percent—constituting almost 21 percent of Americans—agreed either that America always has been and is currently a Christian nation (31 percent) or that the United States is not a Christian nation today but used to be one in the past (41 percent).[84] Both views have been promoted by leaders associated with the Christian Right.[85]

That my focus group participants, drawn from a variety of historically white evangelical denominations in Georgia, matched the traditionalist profile is not a coincidence. Surveys comparing the attitudes of evangelicals from around the country have revealed that southern evangelicals are the most conservative in terms of their voting patterns, views of biblical authority, and attitudes toward social issues. As was borne out in my own research, they are also twice as likely as non-southerners to say they attend church more than once a week.[86] There is little doubt that the South is a stronghold of traditionalism; southern evangelicalism may even have helped drive the broader resurgence of evangelical Protestantism into public life during the last quarter of the twentieth century.[87]

Southern evangelicals may be a unique subset of the broader tradition, but they have nevertheless been remarkably influential. Many leaders in the Religious Right are southerners, and many organizations of the Religious Right are or were headquartered in the South.[88] To a significant extent, southern evangelicals' concerns have driven the broader evangelical agenda and style.[89] Thus, Georgia provided an ideal location to investigate how supporters of the Christian Right viewed climate change.

READER'S GUIDE

The book is organized into two parts that generally follow the path of my research. Part One is based mostly on my qualitative field research and addresses the question of *why* traditionalist evangelicals tend to be climate skeptics. Chapter 1 discusses the origins and evidence in support of the end-time apathy hypothesis, highlighting its widespread acceptance in environmental circles. Chapter 2 describes my informants' basic attitudes toward the environment, as I encountered them in the field. Chapter 3 presents the findings from my field research related to end-time beliefs in particular. Concluding that the end-time apathy hypothesis is conceptually impoverished, chapter 4 presents an explanation that I feel is better grounded in the data: that it is evangelicals' embattled mentality that undergirds and sustains climate skepticism. Chapter 5 further develops this argument, focusing on how skepticism is sustained socially within the religious communities of evangelical churches. Chapter 6 puts this explanation to test via a case study. If it is the embattled mentality that drives climate skepticism, what explains the development of a climate change initiative within the theologically conservative SBC? The answer is, in part, that the Southern Baptists are not a monolithic group. But I also show that many of the climate initiative's high-profile supporters were actually climate skeptics. I argue that they supported it because the embattled mentality led them to an alternative interpretation of the initiative's meaning.

If traditionalist evangelicals' climate skepticism is rooted in their sense of embattlement with secular culture, how did climate change, so conceptually distant from the discourse of embattlement, become part of the narrative? Part Two addresses *how* skepticism came to be established in conservative evangelical circles as the biblical position on climate change. Chapter 7 describes how leaders and pundits associated with the Christian Right conducted a large-scale and at least loosely coordinated campaign to promote climate skepticism in the evangelical mass media, largely without secular observers even being aware that the campaign existed. Lastly, chapter 8 explains why Christian Right leaders were motivated to launch their campaign. It establishes that both evangelical environmentalists and their opponents in the Christian Right had a major stake in convincing lay evangelicals that their position on climate change was the moral, Christian position. The Christian Right, however, was in a better position to disseminate its message to the laity by virtue of its access to mass media and its ability to use trusted messengers and compelling frames.

Thus, the gospel of climate skepticism was born.

Why Traditionalist Evangelicals
Are Climate Skeptics

CHAPTER I

The End-Time Apathy Hypothesis

In 2013, the hip evangelical pastor Mark Driscoll ignited a minor controversy by arguing that there was little need to take care of the environment because Jesus was coming back. "I know who made the environment," he reportedly declared in a speech given at a major evangelical leadership conference. "He's coming back, and he's going to burn it all up. So yes, I drive an SUV."[1] A number of audience members tweeted the line, causing consternation in the blogosphere, where Driscoll was condemned for advocating a "throwaway theology which sees the created world as disposable" and accused of following in Pat Robertson's gaffe-prone footsteps.[2] Was Driscoll representative of a larger trend within evangelicalism?

Blogging for the *Washington Post,* Richard Cizik, an evangelical who had become a major player in evangelical environmental circles via his support of the Evangelical Climate Initiative, seemed to think so. As evidence, he cited a 2011 study showing that 67 percent of white evangelicals believed that the severity of recent natural disasters was evidence of what the Bible called the end times.[3] In the same study, only 52 percent of white evangelicals viewed such events as evidence of global climate change. White evangelicals, in other words, might indeed be unconcerned about climate change in part because they saw it as a sign of the end times. Speaking from his own experience, Cizik added that such views were common: "Those who ascribe to the 'God-will-burn-it-all-down school,' pop up everywhere."[4]

Together with anti-science attitudes and political orientation, end-time beliefs fall among the most commonly proposed popular explanations for evangelicals' lack of environmental concern. But how good of an explanation is it? This chapter describes the origins of the idea and makes a case that our understanding of how end-time beliefs actually produce environmental apathy is limited, having been bolstered more by historical circumstances and inference than by direct evidence, and that further investigation is therefore needed.

THE PERVASIVENESS OF THE END-TIME APATHY HYPOTHESIS

Although the end-time apathy hypothesis is not necessarily well known outside of environmentalist circles, it has come to be a standard explanation for conservative Christians' lack of environmental concern. It has appeared in influential environmental books and magazines, on the lips of well-regarded scientists and politicians, and in numerous scholarly and popular works. In other words, it is not simply the paranoid speculation of environmentalists on the margins, but an idea that is expressed by the most prominent and celebrated among them. A few examples will suffice to illustrate the point.

One of the highest-profile appearances of the end-time apathy hypothesis was in Al Gore's 1992 book *Earth in the Balance: Ecology and the Human Spirit,* published not long before he was elected vice president. In the book, which had sold over half a million copies by 2000, Gore argued that "for some Christians, the prophetic vision of the apocalypse is used—in my view, unforgivably—as an excuse for abdicating their responsibility to be good stewards of God's creation."[5] The role of end-time beliefs in discouraging environmental concern was not a central point of Gore's book, but his reference to it certainly gave the impression that there was no doubt that end-time beliefs were an important driver of environmental apathy.

Another high-profile example of the end-time apathy hypothesis comes from the eminent Harvard biologist and biodiversity advocate E.O. Wilson. In a fictitious "Letter to a Southern Baptist Pastor" that appeared in his 2006 book *The Creation: An Appeal to Save Life on Earth,* Wilson pleaded for scientists and religious practitioners to "set aside our difference in order to save the Creation." Still, he was troubled by a major barrier: the "perplexing" but "widespread conviction among Christians that the Second Coming is imminent, and that therefore the condition of the planet is of little consequence." "For those who believe this form of Christianity," he concluded, "the fate of ten million other life forms indeed does not matter."[6]

Written for a narrower academic audience, the historian Roderick Nash's important book *The Rights of Nature: A History of Environmental Ethics* also suggested that Christianity discouraged environmental concern, in part because

> Christians expected that the earth would not be around for long. A vengeful God would destroy it, and all unredeemed nature, with floods or drought or fire. Obviously this eschatology was a poor basis from which to argue for environmental ethics in any guise. Why take care of what you expected to be obliterated?[7]

Nash is less well known outside of academia than Gore or Wilson, but he is a highly regarded environmental scholar. His reference to the end-time apathy hypothesis further demonstrates the caliber of individuals who have voiced it.

The end-time apathy hypothesis was even included in a popular environmental science textbook, *Biosphere 2000: Protecting Our Global Environment* (2000). In answering the question "Why do uncontrolled population growth, resource abuse and pollution occur?" the authors asserted that Christianity deserved some of the blame because it had helped shape the overarching worldview in the West. According to them, it taught not only that "God gave humans dominion over creation" but also "predicted that God would bring a cataclysmic end to the Earth sometime in the future."[8] "Because these beliefs tended to devalue that natural world," the authors concluded, "they fostered attitudes and behaviors that had a negative effect on the environment."[9] Not only was the end-time apathy hypothesis articulated by educated elites; it was literally the textbook explanation for Christian environmental apathy.

None of these texts argued that end-time beliefs alone were responsible for Christians' environmental attitudes, yet by choosing to highlight this explanation, these works illustrated the pervasiveness of the idea that end-time beliefs are an important factor diminishing Christians' level of concern about the environment.[10] Interestingly, the notion was apparently so obvious that those expressing it felt little need to include supporting evidence. Gore referenced a single individual who had purportedly downplayed the need for environmental protection because of his belief in the apocalypse (James Watt, who is discussed below), while Wilson cited a poll about the prevalence of prophecy belief—a poll that said nothing about environmental attitudes. Nash and the *Biosphere 2000* textbook offered no evidence at all in support of their claims.

So where did this idea originate, and how did it come to be regarded as self-evident?

THE ORIGINS OF THE END-TIME APATHY HYPOTHESIS

In a narrow sense, many Americans were introduced to the possibility that end-time beliefs might cause apathy about the environment in 1981 when it was widely reported that Ronald Reagan's secretary of the Interior, James Watt, had said there was little need to be concerned about conserving the nation's natural resources because "I do not know how many future generations we can count on before the Lord returns."[11] Watt's quotation has echoed through the decades, becoming the most commonly cited example of end-time-inspired environmental apathy. It is the direct origin of the end-time apathy hypothesis.[12] But this quotation alone cannot fully explain the widespread circulation of the hypothesis. First, Watt never actually justified his philosophy of natural resource management with reference to his end-time beliefs. The quotation is real, but the full context in which he uttered it reveals a different intent. But even if it had been interpreted correctly, Watt was just one person. How could one person's views be used to indict an entire faith? I suggest that the answer has to do with what Watt represented.

Watt's high-profile position publicized the end-time apathy hypothesis, but if he had simply been a person who happened to be an evangelical Christian and who also happened to hold an environmentally consequential political appointment, people likely would have forgotten about the end-time apathy hypothesis once it faded from the headlines. The fact was, however, that Watt represented a political shift of tectonic proportions. An evangelical Christian as well as an anti-environmentalist, Watt personally embodied both the ascent of the "New Christian Right" (as the Christian Right was called at the time) and the aggressive anti-environmental backlash movement that were both emerging in the early 1980s. As a closer look at discussions of the environmental impact of end-time beliefs during the Reagan era will reveal, it was this larger political context—in which environmentalists were newly pitted against a Republican party that was both anti-environmental and supported by evangelical Christians—that made the end-time apathy hypothesis seem so plausible. Underlining the importance of political context, the pattern would repeat itself during the George W. Bush administration.

My point in showing that political circumstances contributed to the wide circulation and acceptance (among environmentalists, at least) of the end-time apathy hypothesis is not to assert that it is false. Rather, my point is that we do not know to what extent it is true because its apparent plausibility has for too long made further investigation seem unnecessary.

Watt and the Anti-Environmental Backlash Movement

Reagan had been indifferent to environmental issues as governor of California, but the pro-business, small-government platform of his presidential campaign directly conflicted with environmentalists' strategy during the 1960s and '70s of pursuing environmental protection at the federal level. Reagan made this threat clear when he expressed support for the most important anti-environmental movement of the time, the Sagebrush Rebellion. Sagebrush Rebels, who were based largely in the West, argued that the federal government was an inefficient land manager and should cede control of public lands to state and local officials who could be more sensitive to local needs. To environmentalists and other critics, however, the rebellion was "a land grab, with states fronting for certain interest groups who [sought] exclusive rights to exploit the resources of these lands free from any regulation."[13] The movement was especially threatening to environmentalists because state-owned lands would not be eligible for federal "wilderness" status, a designation that environmentalists counted on to protect natural areas from urban growth and industrial, agricultural, and consumptive recreational uses. Hence, Reagan's statement of support for the movement—"Count me in as a rebel," he assured an audience in Utah while campaigning for president in 1980—was deeply worrisome to environmentalists.[14]

Reagan's pledge of support for the Sagebrush Rebellion became a reality when he appointed Watt, also a self-proclaimed Sagebrush Rebel, to the most environmentally consequential political position of the era, secretary of the Interior. From the perspective of environmentalists, the appointment was a disaster. Watt was "planning to unravel the whole conservation fabric of the country, basically repeal the gains of the 20th century in conservation," warned the executive director of the Sierra Club. "It's appalling."[15]

Environmentalists had good reason to be suspicious of Watt. In 1977, he had been hired as the first president and chief legal officer of a conservative public-interest law firm known as the Mountain States

Legal Foundation (MSLF).[16] In that position, Watt sued (successfully in many cases) the Environmental Protection Agency, the Sierra Club, the Environmental Defense Fund, and the Department of the Interior, mounting forty-seven cases in all.[17] Hence, in a playbook that would become familiar in subsequent decades, in being nominated to the position of secretary of the Interior, he was being called to oversee an organization whose mission he had previously worked to thwart. His transition to secretary of the Interior therefore epitomized the environmental movement's perception that

> in the Reagan years, most of the federal offices responsible for the environment became foxes' dens for profit-making special interests. These political foxes did not have to sneak into the henhouses through a hole in the floor. They were handed the key to the front door and turned loose on the chickens.[18]

As secretary of the Interior, Watt applied Reagan's mission of reducing government regulation to the nation's natural resources and wildlands with zeal. After little more than a year in office, he had enraged environmentalists by proposing to halt the acquisition of new national parks; opening public lands for mineral, oil, and coal extraction; hastening the sale of offshore oil and gas leases; and decommissioning the department's surface mining enforcement program.[19] Environmentalists responded by mounting a petition campaign calling for his resignation, eventually collecting an astounding one million signatures (the signatures had no effect, however). They were enraged enough to demonstrate at his public talks, carrying signs emblazoned with slogans like "Dump Watt" and, for good measure, "Kill a Watt, Save a Tree."[20]

Stationed in the vanguard of the emerging anti-environmental backlash movement, Watt quickly became environmentalists' number one enemy, the emblem of all they hated about the new political order. But Watt was also emblematic of another change: the emergence of the New Christian Right. Watt's dual identity as an anti-environmentalist and a "born-again" Christian—at a time when both were newly in the spotlight—seemed to beg explanation. Before long, this would propel his beliefs regarding biblical prophecy to the foreground.

Watt and the New Christian Right

Reagan's ascent to the White House was notable not only because it reversed the environmental movement's fortunes, but also because it heralded the emergence of the New Christian Right as a political force. The change had been a long time in coming, but it was in 1980 that it finally emerged into public view. Suddenly, evangelicals went from political obscurity—having supposedly opted out of politics in the 1920s after failing in their attempts to keep evolution out of public schools—to center stage.[21] Having actively courted the evangelical vote in his 1980 election campaign, Reagan then appointed two evangelicals to cabinet-level positions as a reward for their support.[22] One of them was James Watt. At a time when evangelicals were newly coming to national attention, observers were primed to pay close attention to how this prominent evangelical appointee combined his faith with public service.

Watt had converted to the Assemblies of God denomination as an adult, and his faith remained, according to his wife, "an integral part of his life" as secretary of the Interior.[23] Environmentalists initially paid little attention to this background, however, focusing their objections to his appointment on his anti-environmental record, well established during his time as president of the MSLF. Thanks to environmentalists' testimony, he attracted more negative votes than any other cabinet member at his confirmation, but he was approved handily nevertheless.[24]

It was only after a few months in office, during which time he rapidly established himself as a foe of environmentalists, that stories noting his religious background began to appear. In February, the *Washington Post* printed a rumor about his conservative taste in women's attire, suggesting that he had ordered female employees to wear skirts.[25] The next month, a front-page profile published in the same newspaper included a description of Watt's religious background and suggested that it caused him to have a "religious" approach to his job.[26] A *Time* magazine profile published that same month emphasized Watt's religious background even more; the piece, titled "Zealous Lord of a Vast Domain," described him as grounding his philosophy of natural resource management in "the self-righteous conviction of the born-again Christian."[27]

It was not until May that observers began to focus on one particular aspect of his religiosity—his end-time beliefs. In a front-page profile published in the *Wall Street Journal,* the headline of which noted his "religious zeal" and "rigid, born-again approach," he was quoted as responding to a question about whether he favored preserving wilderness for future generations by saying, "I do not know how many generations we can count on before the Lord returns."[28] According to the reporter, this statement revealed "a great deal" about Watt's desire to open up America's wildlands for multiple uses—he was, "above all, a devout and supremely self-confident servant of God." The reporter did not say that the reason Watt was so willing to let the nation's natural resources be used up was because he did not think the earth would be around much longer, but others soon interpreted it this way.[29]

The story line caught on immediately, in part because there was a new need to understand the views of these previously invisible millions of Americans, including their beliefs about the end times. As the political scientist Michael Barkun described the situation in 1983, secular academics who were "once prone to ignore or dismiss apocalyptic thought as the artifact of a prescientific age . . . now regard it as simultaneously important, attractive and potentially dangerous."[30] That the Republican Party was now supported by evangelicals, he added, "suggests that to take apocalypticism seriously is prudent."[31]

Hence, the story about Watt's end-time beliefs came to be repeated frequently not only because of Watt's controversial positions on environmental policy, but because it served as a convenient hook for stories about the new political landscape created by the rise of the Christian Right. In June 1982, for example, the *Atlantic Monthly* took on the task of analyzing this new political landscape, publishing an article about the growing popularity of apocalyptic prophecy and its potential political impact. After describing evangelicals' end-time beliefs in lurid detail, the author argued that these beliefs would have real-world effects because the people who held them were now, with the rise of the Christian Right, in positions of power. "Most significant," he suggested, was the possibility that end-time theology would guide these individuals' policymaking. Watt was the example:

> Secretary of the Interior James Watt has emphatically denied that his remark . . . 'I do not know how many future generations we can count on before the Lord returns' implied that we should not be greatly concerned with long-range husbanding of natural resources. But one is hard-pressed to understand why he raised the issue if that was not precisely what he meant.[32]

It was clear from the author's tone that he felt he was introducing a new, hidden world to a broader public. "Alien as it may appear to those unacquainted with it," he wrote, "this interpretation is bedrock and touchstone to millions of fundamentalist Christians." Even if it seemed bizarre, it was not one they could afford to ignore: "Because it is part of a long intellectual tradition with extensive and systematic content, it deserves to be accorded serious examination, not to be dismissed as nonsense," he admonished.[33]

At the *New Yorker*, staff writer Jonathan Schell's worries were similar. For him, James Watt epitomized the "newly political strain of Christian fundamentalism" that Reagan had ushered in. After giving the "Lord returns" quotation, Schell criticized Watt's views as "a novel . . . amalgam of millenarianism and a sort of generational hedonism that has little to do with traditional religious ideas. . . . [B]elieving that the end may be near anyway, [Watt] apparently regards the earth as a disposable item—something to be used up and thrown away."[34] For Schell and other observers of American political life, Watt's justification of policy by faith showed what was in store for America with the emergence of the Christian Right.

For environmentalists, Watt's religious beliefs were an especially provocative explanation for his anti-environmentalism. According to a piece in *Audubon* magazine, for example, the "real source" of Watt's anti-environmentalism was his religious beliefs, which taught that "the Earth is merely a temporary way station on the road to eternal life. . . . The Earth was put here by the Lord for His people to subdue and to use for profitable purposes on their way to the hereafter."[35] In the Sierra Club's *Sierra* magazine, Nathaniel Pryor Reed, a Republican and former assistant secretary of the Department of the Interior, described himself as "sick at heart over the direction his party has been taking in the last few months." He did not blame Reagan, however, whom he believed would be "a notable environmentalist," but Watt, who "seems to think he has divine sanction for his effort to turn our natural resources over to concessioners and developers."[36] And in an impassioned editorial printed in the *Chicago Tribune*, the renowned photographer and environmental activist Ansel Adams decried Watt's religiously based disregard for the future, writing that Watt had "no right to impose his religious philosophy on the management of his department and on the future of the American people."[37]

The story continued to circulate widely, inspiring news coverage in a number of high-profile venues. The *Washington Post* printed an article and an editorial cartoon (figure 1), while one or several articles also appeared in the *Wall Street Journal*, the *Washington Star*, and the *Chicago Tribune*.[38] The story also inspired critical commentary by the influential church historian Martin Marty, and a National Public Radio story in which former secretary of the Interior Cecil Andrus explained that "when we do have the second coming, I do not believe the Lord expects his place to be a barren desert used up by man."[39]

It is always difficult to determine why a particular topic comes to dominate the national conversation, but in the case of Watt, historical context looms large. With his bold, uncompromising style, Watt himself

FIGURE 1. A Herblock cartoon depicting James Watt, printed in the *Washington Post* on June 3, 1981, (c) The Herb Block Foundation. According to Watt's wife, Leilani, the cartoon was devastating to them both. Watt was "fighting tears" after seeing it. For her part, Leilani "felt as if a sword had been run through me." She wondered, "How could anyone so misrepresent my husband's Christian faith or the Interior policies?" (Watt, *Caught in the Conflict*, 91–92).

did much to chisel the image of the environmentally rapacious end-time believer in the nation's collective conscience. But it was the simultaneous emergence of the Christian Right and the anti-environmental backlash movement that made the image enduring and powerful. For observers flummoxed by their simultaneous rise, the end-time apathy hypothesis provided a simple but powerful explanatory framework. Continuing to follow the story of the end-time apathy hypothesis will show that it did not fade from mind, but in fact returned to center stage during the era of growing concern about climate change.

End-Time Beliefs and Climate Policy in the George W. Bush Era

Like Reagan before him, George W. Bush sought the support of evangelicals. But unlike Reagan, who had come from a mainline denomination, Bush was himself an evangelical, having converted as an adult in order to overcome a drinking problem.[40] Having actively courted the evangelical vote during his election campaign, Bush went on to champion a number of evangelical causes while in office. As the political scientists Kenneth Wald and Allison Calhoun-Brown relate, during the Bush administration "religious conservatives enjoyed unparalleled access to the levers of power."[41] As in the Reagan era, this raised questions about the role and motives of evangelical Christians involved in government.

In another parallel with Reagan, Bush was also widely despised by environmentalists for his environmental record. According to John Dorner, a spokesman for the Sierra Club, during his two terms in office Bush "[undid] decades if not a century of progress on the environment."[42] Especially unforgivable in the eyes of environmentalists was Bush's decision not to implement the Kyoto Protocol, the first major international agreement designed to tackle climate change. The Bush administration not only refused to participate in international efforts to stabilize the climate, but sought to justify this inaction through techniques "ranging from emphasizing the 'uncertainty' of climate science . . . to suppressing the work of government scientists."[43] Many environmentalists viewed such techniques as intentionally obstructive and deceptive. Beyond blocking action on climate change, the Bush administration also upset environmentalists by taking such actions as dismantling the protections of the Endangered Species Act, amending the Clean Air and Clean Water acts to reduce the protections they offered, and opening millions of acres of designated, publicly owned wilderness to mining, oil and gas drilling, and logging.[44]

As had been the case during Reagan's time in office, the Bush administration's anti-environmental posture led some environmentalists to begin speculating about the administration's links with conservative Christianity. Days after the opening of the 108th congressional session in 2003, a freelance journalist named Glenn Scherer wrote a story for the news and opinion website *Salon* that blamed the Bush administration's assault on nature not only on economic and political motivations, but on "a more deep-seated cause, one tougher for the secular mind to grasp": the Christian faith. According to Scherer, "many fundamentalists see dying coral reefs, melting ice caps and other environmental destruction not as an urgent call to action but as God's will." Going further, he claimed that "within the religious right worldview, the wreck of the earth is Good News!"[45]

Scherer's article was just one of many published about the administration's environmental policies at the time, but it gained some prominence when it was republished on the website AlterNet and, later, as a cover story in *E-the Environmental Magazine*.[46] Both venues cater to progressive, environmentally concerned audiences, so the stories helped to spread (or reinforce, for those familiar with the Watt story) the idea that the administration's anti-environmentalism was motivated in part by the end-time beliefs of its Christian leaders. The only change was that the focus had shifted from how end-time beliefs might impact public lands to how they might impact climate policy.

Scherer soon wrote a second article that focused even more sharply on the role of end-time beliefs, published in the online environmental magazine *Grist* and originally titled "The Godly Must Be Crazy."[47] A few scholars, most notably Paul Boyer (whom Scherer cited), Daniel Wojcik, and Michael Barkun had already described the tendency of religious apocalyptics to consider environmental disaster a sign of the coming end times.[48] But the timeliness of Scherer's piece and the popular venue in which it was published meant that it reached a much larger audience. Garnering thirty-one thousand views in the first month it was online, it was *Grist*'s most popular story that year.[49]

Part of the article's popularity had to do with how well the argument spoke to the anxieties that this era (like the Reagan era) had provoked in many liberal and secular Americans. At the same time that best-selling books like Kevin Phillips's *American Theocracy* (2006) were decrying the intrusion of conservative Christianity into politics in general, Scherer suggested that the phenomenon had sobering implications for environmental policy in particular.[50] Like Phillips's book, Scherer's article sought to create "a harrowing picture of national danger that no American reader will welcome, but that none should ignore."[51]

A story like Scherer's could easily have remained on the margins, but instead it was spread further via the journalist and cultural/political commentator Bill Moyers. In his acceptance speech for the Harvard Center for Health and the Global Environment's "Global Environmental Citizen Award," given in December 2004, Moyers synthesized

Scherer's work with that of other scholars and journalists to argue, like Scherer, that the ascendance of fundamentalists to political power was a disaster for the environment. In his words, "A powerful current connects the [Bush] administration's multinational corporate cronies who regard the environment as ripe for the picking and a hard-core constituency of fundamentalists who regard the environment as fuel for the fire that is coming." Nodding to the salacious appeal of the end-time apathy hypothesis, he noted that the idea of end-time-related environmental apathy "sends a shiver down my spine."[52]

Moyers's speech also hinted at why Watt's reputation as an environmentally apathetic end-time believer had only grown over time. Both Moyers and Scherer had used Watt to illustrate their thesis, giving an apocryphal quotation that had long been incorrectly attributed to Watt: "After the last tree is felled, Christ will come back." In apologizing for and retracting the quotation, Moyers observed that "it is difficult in this cyberworld to catch up with an error. Once something like this begins to circulate, it takes on a life of its own."[53]

The speech attracted considerable attention, especially in the liberal and environmental media. It became a cover story for the *New York Review of Books,* an influential literary-intellectual magazine,[54] which also published the speech in book form the next year.[55] Beyond being posted on Moyers's website, it also circulated widely on the Internet and was excerpted in newspapers and magazines like the *Miami Herald,* the *Ecologist,* and the *Oakland Tribune.*[56] Numerous other regional newspapers reprinted an editorial that presented a brief, approving digest of Moyers's argument.[57] Such stories ensured that the end-time apathy hypothesis was spread far and wide, much as it had been during the Reagan administration. Meanwhile, the convergence of anti-environmentalism and evangelical political activism under way during George W. Bush's time in office seemed to make a strong circumstantial case for its validity. But was it actually true?

ENVIRONMENTALLY APATHETIC END-TIME BELIEVERS

Three types of evidence support the end-time apathy hypothesis. The first is anecdotal: stories of individuals who do not care about the fate of the environment because of their end-time beliefs. Second, there are the gut convictions of some evangelicals or former evangelicals that end-time beliefs diminish environmental concern. Finally, there are social scientific studies testing the strength of the association between end-time beliefs and environmental apathy. Examining each in turn will show what is known about how end-time beliefs affect environmental attitudes, as well as where the picture is incomplete.

Anecdotes play a special role in the literature articulating the end-time apathy hypothesis because they appear to provide definitive proof of a connection between end-time beliefs and environmental apathy. In terms of evidence, they are the proverbial "smoking gun." While social scientific studies can demonstrate a correlation between end-time beliefs and environmental apathy, it is harder to prove that end-time beliefs are actually what is *causing* the apathy, because the observed correlation could (in theory) be caused by another aspect of faith connected to end-time beliefs. Hence, it is the existence of individuals who articulate these views that lends credence to the theory that end-time beliefs are, in fact, to blame. Since no gun was more smoking than James Watt, it is worth looking closely at how this apparent case of end-time-inspired environmental apathy unravels.

The original source of the story about his end-time-inspired environmental indifference was a February orientation briefing to the House Interior and Insular Affairs Committee. There, Representative James Weaver of Oregon had asked Watt whether he agreed with "the general statement that we should save some of our resources . . . for our children."[58] Watt replied:

> Absolutely. That is the delicate balance the Secretary of the Interior must have, to be steward for the natural resources for this generation as well as future generations. I do not know how many future generations we can count on before the Lord returns. Whatever it is, we have to manage with a skill to leave the resources for future generations.[59]

Despite how it was later interpreted, his full original statement clearly expressed a commitment to steward the earth's natural resources.[60]

Watt attempted to correct the record at various points, but his answers never satisfied those who suspected that his end-time beliefs played a covert role in his philosophy of natural resource management. For example, in a House oversight hearing on the reorganization of the Office of Surface Mining, held about six months after the February orientation, Representative Weaver again questioned Watt. Even though he had heard the latter's full original statement at the earlier briefing, he pressed Watt repeatedly. In a contentious exchange, Weaver asked him, "Are you approaching the environmental issue of surface mining [with] 'Why worry, the Lord's return is imminent?'" Watt first tried to turn the conversation back to his professional duties, but when Weaver continued to press him, Watt replied, "I am tired of people telling me what I have said. I never said that." After reminding Weaver that he had heard the original statement, Watt again attempted to halt the line of questioning, this time asserting that "my religious freedom is guaranteed under the First Amendment. I don't think that it needs to be an issue of this hearing." When another Republican senator agreed that Weaver's line of questioning was "completely out of line," the conversation moved on—likely to no one's satisfaction.[61]

Some observers may have interpreted Watt's refusal to discuss the matter further as an admission of guilt, but according to his wife, Leilani, during her husband's time in office he generally refused to discuss his religious beliefs because he believed that God would defend him, and because he felt that the press would distort whatever he said.[62] He did state in a 1981 interview that "it's been 2,000 years since the last coming of Christ and it might be another 2,000 before the second coming."[63] By that point, however, it was too late to stop the story's spread.

The widespread news coverage and frequent appearance of the shortened quotation left an indelible mark not only on Watt's reputation but on the environmental reputation of evangelicals more broadly. As David Kenneth Larsen concluded in his intellectual history of the evangelical environmental movement, Lynn White Jr. had popularized the notion that Christianity was responsible for the environmental crisis, but it was James Watt who convinced environmentalists to "single out those with 'a very literal interpretation of Scripture' as being especially culpable for environmental degradation."[64]

When I investigated other anecdotes describing end-time-inspired environmental apathy, I found to my surprise that a number of them unraveled in a similar manner.[65] To take one example, Mark Driscoll, the evangelical leader discussed at the opening of the chapter, was criticized for saying it was permissible to drive an SUV because Jesus was going to "burn it all up" anyway. But further investigation suggests that, similar to Watt's, his words had been taken out of context. As Driscoll explained later in a blog post, he had been giving a talk about how leaders should find their identity in Jesus rather than in their possessions or professions. He had relied on jokes to underline the point, including a series about how transportation choices reflected identity. "If I remember correctly," he wrote, "this segment included jokes about hipsters who ride scooters, truck dudes, minivans driven by guys who feel like a mini-man (notice the clever combination), and SUVs driven by people who do not care about the environment." In context, the joke was actually *on* people who did not care about the environment, rather than an endorsement of anti-environmental practices.

Driscoll even considered himself an environmentalist, despite being someone who believed that Jesus would personally return to earth someday. "For the record, I really like this planet," he stated. "We should take good care of this planet until [Jesus] comes back to make a new earth, like the Bible says he will." If we can believe his account—and it seems reasonable to do so, given that he has not publicly reversed his views on controversial issues like traditional gender roles or the sinfulness of homosexuality—Driscoll provided evidence *against* the end-time apathy hypothesis, not for it.[66]

The apocryphal origins of end-time-related anecdotes make it especially clear that further evidence is needed. Still, the convictions of insiders gave me pause. Who would be in a better position to know whether end-time beliefs really inspire environmental apathy than the people sitting in the pews every week? It is particularly hard to dismiss the speculations of E. O. Wilson, Al Gore, and Bill Moyers, given that they all were raised as evangelicals.[67] Surely, they would not have made the claim if they had not seen some evidence.[68]

Such intuitions must be weighed, however, against the statements of other evangelicals who counter them. Watt, for one, excoriated Moyers for using him as an example of end-time-inspired apathy. In a vituperative op-ed published in the *Washington Post,* he wrote that "I never said it. Never believed it. Never even thought it. I know no Christian who believes or preaches such error. The Bible commands conservation—that we as Christians be careful stewards of the land and resources entrusted to us by the Creator."[69] Leaving no doubt as to his thoughts on the legitimacy of the end-time apathy hypothesis, he concluded that "if such a body of belief exists, I would totally reject it, as would all of my friends."[70]

Offering a slightly different take, in 2005 the EEN's executive director, Jim Ball, wrote in an online editorial that he accepted that end-time beliefs reduced environmental concern for some Christians, but "end-times views are a half-mile wide . . . but only an inch deep. Hardly anyone makes important day-to-day decisions based upon these end-times views."[71] Such disputes indicate that a more systematic approach is needed.

Turning to the social scientific literature, the first question one might ask is whether surveys have, in fact, found end-time beliefs to be associated with lower levels of environmental concern. Finding an association would give us more confidence that the anecdotal evidence and gut convictions are pointing to something real and significant—something that affects not just a few people, but a wide range. Moreover, we would want to know whether this association persists after controls are added. To draw on the other major theories about the reason for evangelicals' lower levels of environmental concern, could the association between end-time beliefs and lower levels of environmental concern in fact be caused by end-time believers being more politically conservative, more skeptical of science, favoring action on the individual rather than collective level, holding dominion beliefs, or some other factor associated with their faith?

The handful of social scientific studies that have focused on end-time beliefs cannot satisfy our curiosity about all of these questions, but they do clearly show that end-time beliefs are associated with lower levels of environmental concern even after political and demographic controls are added. A study that sampled members of eight religious interest groups (ranging from the politically liberal Evangelicals for Social Action to the politically conservative Concerned Women for America) in the early 1990s, for example, found that biblical literalism, social pessimism, and end-time thinking were all associated with reduced environmental concern among fundamentalists, even after political and demographic controls were introduced.[72] In terms of comparative influence, politics initially appeared to be more important than religion, but because many theologically conservative Protestants are politically conservative *because of* their theological commitments, the authors ultimately concluded that religious factors had a greater impact on environmental concern than political ideology.[73]

The study just cited was unable to isolate the effects of end-time beliefs specifically, so the same authors conducted a second study to examine their effects more closely. This time the group found that end-time beliefs were "by far the strongest religious predictor of environmental perspectives"—having an even stronger effect in most of their samples than political affiliation.[74] Interestingly, in this study the association between end-time thinking and lack of environmental concern was particularly strong among two of their samples: religious activists (which included both laypeople and clergy) and the activist public. The religious activist sample consisted of respondents who belonged to religious interest groups, including several associated with the Christian Right: Focus on the Family, the Prison Fellowship, and Concerned Women for America. The second sample was composed of the top 15 percent of citizens in terms of political involvement. In other words, this study's data not only showed that the connection between end-time belief and lack of environmental concern was statistically significant after controlling for other factors, but that this effect was the strongest among the politically mobilized.

A third study on end-time beliefs and environmental attitudes, published in 2012, is of particular interest because it focused on the effect of end-time beliefs on attitudes toward climate change. The results showed that the effect identified in the 1990s had persisted into the next decade. After controlling for biblical literalism, education, and demographic factors, the political scientists David Barker and David Bearce found that belief in the Second Coming reduced the probability of strongly agreeing that the government should take action to address climate change by 12 percent.[75] When party identification, ideology, and media distrust were removed from the model—because these were "to some extent theoretically subsequent to religious beliefs"—that effect increased to 20 percent.[76] This made religion a serious impediment to action, the authors concluded, because "the fact that such an overwhelming percentage of *Republican* citizens profess a belief in the Second Coming (76 percent in 2006, according to our sample) suggests that government attempts to curb greenhouse emissions would encounter stiff resistance even if every Democrat in the country wanted to curb them."[77]

Finally, a study published in 2018 found that evangelicals were, at a statistically significant level, less likely than mainline Protestants to be concerned about some future environmental problems. Two of the three future environmental problems that the authors tested demonstrated this relationship, a finding that the authors attributed to the impact of eschatology (citing the studies just mentioned, since they did not measure eschatology directly).[78]

Collectively, these studies indicate that the gut convictions discussed above were not entirely misplaced: end-time beliefs are associated with reduced environmental concern, even after controlling for other possible influences. They do leave certain questions unanswered, however. Most importantly, they use James Watt and just a few others to illustrate what the association between end-time beliefs and environmental apathy looks like. Yet removing Watt from the picture puts enormous interpretive weight on just a few individuals, who are assumed to represent how millions of evangelicals understand environmental decline. What is it actually like to be so convinced the end is near that one does not care about the state of the planet? Do such beliefs simply discourage people from thinking in the long term or, to entertain the most disturbing possibility, do they actually make people excited, rather than dismayed, about signs of environmental degradation? Are there people out there who actually feel energized by news of deforestation, species loss, and abnormally warm winters? And why would a supposedly otherworldly theological belief have the strongest effect among politically engaged evangelicals—that is, among those with the sharpest focus on this-worldly activity? How exactly, in other words, did end-time thinking produce environmental apathy?

A STRATEGY FOR UNDERSTANDING HOW END-TIME BELIEFS AFFECT ENVIRONMENTAL ATTITUDES

While I generally wanted to understand why so many social scientific studies had found theologically conservative Christians to be unconcerned about the environment compared to other Americans, my interest in end-time beliefs dictated a more focused research strategy. I decided, therefore, to divide my focus groups into two groups based on their beliefs about the end times.

Understanding this division will require a brief discussion of end-time beliefs in Christianity. When scholars, journalists, and other observers talk about end-time beliefs in the context of environmental concern, they typically single out for blame a school of eschatological interpretation known as premillennialism. But premillennialism is actually one of three types of Christian thinking about the end times.[79] These schools differ according to where adherents place the messiah's return in relation to the thousand-year period (a millennium) of peace and righteousness that will accompany that return. *Premillennialists* believe that Jesus will return before the millennium; *postmillennialists* believe that he will return afterward; and *amillennialists* "interpret biblical references to the millennium figuratively and contend that the millennial reign of Christ occurs in the hearts of his followers."[80]

The social implications of these different eschatological schools have changed over time, but in the modern context, premillennialism is the school that is associated with a social pessimism and interest in signs of the nearing end (e.g., wars and rumors of war, natural disasters, moral decadence, and increasing apostasy).[81] By contrast, as one textbook on Christian millennialism notes, amillennialists "are not preoccupied with biblical prophecy and the signs of the times and have little to say about the deterioration of world conditions and the present culture."[82] Hence they are thought to have a very different outlook on culture than premillennialists.

Regarding who qualifies as an "end-time believer," both premillennialists and postmillennialists of the theologically conservative variety could technically be placed in this category in the sense in which the term is commonly used—to refer to people who believe that Jesus will literally return to earth one day.[83] But in reality, premillennialists are the face of end-time belief in America because premillennialism has become the eschatological school of choice for America's largest group of theological conservatives, evangelicals.

The popularity of premillennialism among American evangelicals today can be traced back to the efforts of a British revivalist named John Nelson Darby who toured America in the late 1800s promoting a type of premillennialism known as dispensationalism. Premillennialism spread throughout the twentieth century, growing via its inclusion in popular reference works like the *Scofield Reference Bible* and in the curricula of influential Bible institutes.[84] Its popularity broadened even further when historical events seemed to confirm biblical prophecies. Particularly key in this regard were the invention of the atomic bomb in 1945 (which provided a mechanism for the world to end in fire, as prophesied in 2 Pet. 3:10), the formation of the state of Israel in 1948 (a precursor to the "gathering of the Jews" thought to occur prior to the Rapture), the advent of the Cold War (Russia was thought to be the menacing "king of the North" that would invade Israel in the last days), and the surge of environmental prophecies of doom in the 1960s and '70s (lending credibility to notions of a world on the brink of disaster).[85]

Evangelicals' adept use of print media also helped popularize end-time beliefs. In the 1970s, Hal Lindsey—a former tugboat captain turned minister for Campus Crusade for Christ—wrote two books that became surprise best sellers, *The Late Great Planet Earth* (1970) and *There's a New World Coming* (1973). Lindsey essentially repeated the premillennialist views he had learned in seminary, but he disseminated these views using contemporary publishing formats and distribution techniques, which greatly increased the number of people they reached.[86] In 1979, *The Late Great Planet Earth* reached an even larger audience when it was turned into a popular film narrated by Orson Welles.[87] Meanwhile, well-known preachers like Billy Graham, Jerry Falwell, Jack Van Impe, Pat Robertson, and Oral Roberts all helped spread end-time beliefs to popular audiences via film, television, and radio. Tim LaHaye's *Left Behind* series of novels, published from 1995 to 2007, continued Lindsey's legacy and helped attract further recruits. With a 2010 poll showing that 58 percent of white evangelicals affirmed that Jesus will return

TABLE 1   DENOMINATIONS AND CHARACTERISTICS OF FOCUS GROUP CHURCHES

Denomination	Denominational family[a]	Setting[b]	Date of focus group	Average weekly service attendance[c]
Premillennialist				
Independent Baptist	Fundamentalist	Rural	August 7, 2011	95
Missionary Baptist	Fundamentalist	Rural	October 2, 2011	20
Assemblies of God	Pentecostal	Semi-urban	October 12, 2011	115
Seventh-day Adventist	Adventist	Suburban	January 28, 2012	120
Amillennialist				
Churches of Christ	Restorationist	Semi-urban	July 17, 2011	70
Presbyterian Church in America	Presbyterian	Semi-urban	November 6, 2011	100
Christian Church	Nondenominational	Semi-urban	October 30, 2011	90
Lutheran, Missouri Synod	Lutheran	Urban	January 18, 2012	220
Southern Baptist[d]	Baptist	Urban	January 22, 2012	550

[a] Pew Research Center, "Religious Landscape Study."

[b] "Semi-urban" refers to cities with a population less than 50,000. "Urban" refers to churches in cities with a population greater than 50,000. "Suburban" refers to churches located in suburbs of either urban or semi-urban areas.

[c] Estimates were based on observation of a randomly selected non-holiday Sunday service (or Saturday for Seventh-day Adventists) and were confirmed by conversations with pastors.

[d] The Southern Baptist Convention's statement of faith does not specify a premillennialist or amillennialist orientation, stating simply that "God, in His own time and in His own way, will bring the world to its appropriate end" (SBC, "The Baptist Faith and Message"). The lack of specificity initially led me to categorize this denomination as amillennialist. The Southern Baptist pastor I interviewed, however, told me the denomination was premillennialist. Subsequent research indicated that Southern Baptists hold a variety of views (Roach, "End Times").

by 2050, it appears that end-time beliefs have continued to be a mainstay of American evangelicalism well into the twenty-first century.[88]

If end-time beliefs really were a major factor reducing environmental concern, I reasoned, it should be possible to discern a difference between the environmental attitudes of premillennialists (presumably unconcerned because of their belief that Jesus would return soon) from those of amillennialists (less likely to be apathetic, in theory, because of their figurative interpretation of the millennium).[89] I therefore arranged for half of my focus groups to occur at evangelical churches that espouse premillennialist views on the end times, and half to occur at evangelical churches that espouse amillennialism (table 1; see appendix A for further details). Eager to discover what the "burn-it-all-downers" really thought about humanity's responsibilities regarding the natural world, I headed to the field.[90]

CHAPTER 2

Practical Environmentalism

By the time I moved to Georgia in the late spring of 2011, it was shaping up to be an unusual year, weather-wise. It had been the most active April for tornadoes in U.S. history, resulting in several deadly events that were widely reported in the national news.[1] Record snow and rainfall in the Midwest over the winter resulted in devastating spring floods along the Mississippi, while the opposite condition prevailed in Georgia. In June, Governor Nathan Deal called for twenty-two counties (including several that I worked in) to be declared federal disaster zones due to a drought that was then in its sixth year. A *New York Times* article about the drought, which affected thirteen other southern states, included a picture of a flat reddish plain, ostensibly a field of cotton, in which only a few stunted seeds had managed to germinate.[2] The drought had so exacerbated conditions that a fire started in April in the Okefenokee National Wildlife Refuge (which straddles the Florida-Georgia border) had consumed nearly two hundred thousand acres of swampland by mid-June. I would skirt the blaze when returning from trips to Florida, and still be able to smell its smoky scent nearly a hundred miles to the north. The blaze would not be put out for a full year.[3]

To be attuned to the environment and, especially, to climate change, is to notice such things with alarm. One of the first things I realized after moving to Georgia, however, was that my informants did not live in the same unsettling world. Their attention was not generally focused on environmental problems at all, and when I brought such problems up in conversation, they usually found ways to redirect the conversation else-where—often back to faith. Hence, to begin to describe their environ-mental attitudes is to describe an absence: absence of interest and absence of concern. Such attitudes did not exactly constitute apathy, however, for they were attached to strong opinions about how the envi-ronment should be treated. Advocating what I call "practical environ-mentalism," they maintained that human treatment of the natural world should be governed by common sense, in keeping with local mores, apolitical, enacted at the individual level, locally scaled, and proudly anthropocentric. Furthermore, while they did agree that certain practices were wrong, they did not express any anxiousness or urgency about environmental issues. Worrying about the environment or focus-ing on environmental problems as problems in and of themselves would indicate a lack of faith in God, and a subversion of the true purpose of life on earth: to seek redemption from sin and thereby attain salvation. In this context, "doing what you can do"—not being wasteful or pur-posefully harming the environment—was their ideal.

This observation begins to shed light on the many studies that have shown conservative Protestants to be less concerned about the environ-ment than other Americans. My informants' evident concern over local issues belies the assumption that they were fundamentally unconcerned about the health of the natural world.[4] Rather, some of what appeared to be environmental indifference resulted from the poor fit between the kinds of approaches and concerns that environmentalists tend to emphasize and my informants' religious and cultural mores. Most prominently, mainstream environmentalism is predicated on a belief in the necessity of activism (which many of my informants viewed as socially disruptive), the need for structural or global solutions (which went against my informants' preference for individual or local-level solutions, leaving global problems to God), and the virtue of contem-plative and/or nonconsumptive interaction with the natural world (while my informants tended most often to speak of their interactions with the environment in terms of consumptive practices like hunting).[5] This is not to suggest that practical environmentalism is an adequate response to today's environmental problems. But it is to recognize something important: even the highly devout, politically conservative, and (as the next chapter explores) end-time-believing Christians I met in Georgia believed in caring for the earth.

THE INVISIBLE ENVIRONMENT

In *A Sand County Almanac,* the legendary conservationist Aldo Leopold hauntingly observed that "one of the penalties of an ecological education is that one lives alone in a world of wounds. Much of the damage inflicted on land is quite invisible to laymen."[6] More than half a century and thousands of books, reports, and newspaper articles later, it is tempting to think that we are now all aware of the earth's wounds. Yet for many Americans, environmental problems remain difficult to perceive because they accumulate slowly, because they become problems only at a scale greater than the individual can access through personal experience, or because ecologically degraded states do not always appear to the untrained eye to be the result of human alteration. As in Leopold's day, perceiving them requires effort and training.

It is because of these perceptual challenges that many environmentalists have worked to render these problems visible. For environmental educator David Orr, for example, cultivating "ecological literacy"—defined as the "capacity to observe nature with insight"—should be a central purpose of education.[7] Similarly, for Mitchell Thomashow, an environmental studies scholar and activist, to confront global environmental change one must develop "perceptual ecology," a set of observational skills and practices that allow one to become grounded in his or her local environment.[8] Only by doing so can one develop "biospheric perception," which helps make sense of the magnitude of changes now occurring at the global scale.[9]

Like many Americans, however, the evangelicals I met had not cultivated such skills. Thus, the first step toward understanding their views about the environment is understanding that the environment was not a particularly salient issue for them. This lack of salience was apparent in many ways. Most obviously, when I asked my focus group informants about environmental issues, I sometimes got curious responses. During a focus group at an Assembly of God church in Mayfield, for example, I asked those present if they had any concerns about the quality of their environment. One woman mentioned that Mayfield had little access to quality consumer goods—Walmart was the only place to shop. When I clarified that I meant the natural environment, the group seemed uncertain how to respond. After a moment of silence, Andrea, a woman in her sixties who had previously been quite talkative, chuckled apologetically that she was "trying to think of something to say." Was I referring to how smoky it was that day?

Eventually the group started to understand what I was asking, but the moment of befuddlement was illuminating. The group had already discussed, articulately and at length, the contributions of Christians to society and the challenges they faced, but they had little to say about environmental problems. When I later pressed them to talk about which environmental problems they had heard the most about, the smoke from Okefenokee fires came up, as did smog, but I could elicit no further examples. In this group, at least, participants could hardly think of examples of environmental problems. This suggests that they were untrained in ecological perception—at least as it is conceived by environmentalists.

The examples that did come up in other focus groups furthered the impression that my informants were not particularly familiar with widely publicized environmental issues such as habitat loss, pollution, threats to biodiversity, and the like (climate change, which I address in later chapters, was something they had heard about, although it was not a top concern either). Instead of referring to such problems, the most common example my informants gave was a recent local fish kill that had resulted from chemical contamination. Even this example came up in only three of the nine focus groups. Other examples were inconsistent and not plentiful, further highlighting the general lack of attention to environmental issues. Nor did Mayfield (a small town, to be sure) have any environmental organizations. According to Jill, the environmental educator I had befriended, students received little such education in the school system either. "The education is about farming. It's an agricultural education," she told me.

My informants were not unique in their lack of attention to environmental issues. As the sociologist Kari Norgaard showed in her ethnography of a small town in Norway, even in a country where climate change was already having a substantial impact, a variety of factors discouraged residents from paying attention to it or deciding to address it.[10] Similarly, focus group research conducted in five cities in New Mexico found that the environment was considered a major issue only among residents of the politically liberal town of Santa Fe.[11] This reflects a broader pattern within the American public, which consistently ranks the environment and climate change as low- to mid-level priorities.[12] Yet my informants in Georgia deflected attention from environmental problems in unique ways, indicating that their cultural and religious beliefs shaped their environmental attitudes at a deep level.

COMMON SENSE AND STAYING OFF THE SOAPBOX

At the most fundamental level, my informants' vision for ideal environmental behavior was grounded in a deep appreciation for common sense. In the words of Joshua, a pastor from the Churches of Christ denomination:

> Without even bringing in religion, I'll say of course we should not litter, put pollution into the air that causes acid rain, of course we should not put nuclear waste in neighborhoods. It's not an issue of right or wrong and sin, it's an issue of common sense and rationality.

This commonsense approach was common to many other informants, who argued that humans (and especially Christians) should not recklessly destroy or damage forests, intentionally pollute, engage in needlessly wasteful behavior, or otherwise wantonly despoil the natural world. "I think the earth is not ours . . . the earth belongs to God and Jesus," shared Kara, a nurse in her forties who attended a Presbyterian church. "And it's just like if I borrow something from a friend," she added, "I am not going to . . . mess it up and then give it back to him. I'm going to make sure whatever I give back to him is in at least as good condition as when [he] let me borrow it."

Joshua described his views as having nothing to do with religion, but the scholarship on evangelicalism suggests otherwise. In particular, my informants' emphasis on common sense closely parallels evangelicals' general approach to biblical interpretation, which also emphasizes common sense over erudition or training. Rooted in the Protestant Reformation's emphasis on the priesthood of all believers, contemporary evangelicals believe that by virtue of God's grace, Christians can understand scripture on their own, without the mediating presence of the church.[13] In a similar fashion, my informants felt that taking care of the environment did not require the intervention of experts, but rather the application of common sense.

There was also a strong sense that environmental behaviors needed to be in keeping with local mores, that they should stay away from extremism and avoid disrupting the status quo. This ideal came through most strongly at the Independent Baptist church, which was located in a rural area outside of Mayfield. When talking about how one should take care of the environment, Rhonda (a school librarian in her fifties) and Michelle (a teacher in her thirties) both emphasized moderation:

> *Rhonda*: We recycle, we do recycle. . . . [And] people will pick up, and we know enough here to take care of what . . . God entrusted us to have. I have a very strong feeling that I must take care of things and I must take care of the environment and the world and not hurt it. But I think you can be extreme.

> *Michelle*: I don't think many people in this area are vocal about it. We just do it. If it needs to be cleaned up you just do it. If it needs to be taken care of, you just do it. You don't have to tell somebody and complain about it and make a big deal about it. You just get off your butt and do it.

Rhonda and Michelle defined environmental behavior as modest in scope: recycling, not littering, and taking care of what they had. Their examples essentially extend the principles of good housekeeping beyond the walls of the house. In the same way that one should maintain a clean, orderly home, one should maintain a clean, orderly environment.

The emphasis on orderliness as a personal and environmental ideal seemed to be a broader regional phenomenon. I once remarked to Jill about how kempt all the churches in the region were, with their fresh coats of white paint and trimmed hedges (they stood out in particular because so many country roads and neighborhoods were blighted by vacant, crumbling houses). She felt that the attention to orderliness in general was related to religion, that keeping the yard trimmed (for example) was a way of demonstrating strong values and an upkeep in faith. In a symbolic fashion, the outside of the house was presumed to reflect the spiritual order that prevailed inside.

Returning to Rhonda's and Michelle's comments, the logic of good housekeeping contained an important implication: much in the same way that one should not tell someone else how to keep his or her house, one should not be "vocal" about telling others how to maintain their environment. In their view, not "making a big deal about" things—neither trying to change people's minds nor change the way they currently did things—was the appropriate course of action. Rhonda further emphasized her preference for moderate environmental behaviors later in the conversation, stating that she brought her own bags to the grocery store and chose paper over plastic if she had forgotten her own. Yet she added that "there are things that I do, but I am not going to change any other things that I'm doing, and I'm not going to stand on a soapbox and say, 'Oh my God the earth is going to die because I use plastic bags.'" Her insistence that she would neither go out of her way nor "stand on a soapbox" captures her vision of a non-activist, moderate environmentalism.

Several of the environmentalists I interviewed corroborated this aversion to disrupting the status quo. Jim led the local chapter of a national environmental organization I will call the National Environmental Coalition (NEC), which was based in Dixon. Born and raised in Georgia, he felt that social conservatism was a major barrier:

> I've heard lots of people say, "I don't believe in environmentalism." And I always say, "You don't believe that you should try to take care of the environment that you live in?" And they'll say, "Well, of course, I think you should take care of the environment you live in. But I think you don't have to argue with the paper plant to do that." And I say, "Well, you do, actually."

In Jim's experience people *did* believe they should protect the local environment ("Well, of course"—it was common sense), but they did not want to battle established forces to do so. Kara, a volunteer with the NEC, had noticed the same reluctance to disrupt the status quo. In her experience, the mentality of longtime residents of her city was that "you don't want to shake things up. You don't want to say things that are going to embarrass anybody or go against the way of doing things." Hence, the first pillar of my informants' practical environmentalism was employing common sense and propriety to maintain a clean, orderly environment.

INDIVIDUALISM AND ANTI-STRUCTURALISM

My informants also displayed a preference for environmental behaviors that were undertaken by individuals (rather than groups) and private (rather than political). This dovetailed with their social conservatism, since private behaviors undertaken by individuals are less likely to disrupt established social norms. It also went along with a specifically evangelical tendency to assume that social problems stem from the sum of individual choices rather than from systems or structures that incentivize bad behavior.

These preferences tended to become apparent when informants were discussing appropriate environmental concern. For example, Lee, a general contractor in his twenties, said he disliked the term *environmentalist* and preferred the terms *naturalist* and *conservationist,* which he defined as

> somebody who [doesn't] want to go dump pollutants in the water, somebody who wants to throw the small fish back or keep the deer population going, or whatever may be. Recycling. Just somebody who's trying to, if you clearcut two hundred acres, you go back and you plant two hundred acres

somewhere else. You're trying to keep the earth going. But they're not out there trying to be Al Gore, or whoever. Not to call him out, but [for a naturalist/conservationist] it's not extreme, it's not a moneymaker.

Here Lee articulated a preference for private behavior over public behavior and for socially conservative over socially disruptive behavior. Regarding the first distinction, the examples he gave are all actions that can be taken in private. Furthermore, whether one is throwing the small fish back or replanting trees, the behaviors he identified all can be performed on the individual level, without any collective action (notably, they also all relate to consumptive uses of the environment, via fishing, hunting, or logging). Finally, because they are private, individual behaviors, they would not disrupt any established social norms, nor would they require becoming politically engaged.

According to sociologists Michael Emerson and Christian Smith, the focus on individual-level behaviors and relationships rather than structural ones is part of a broader habit of mind among white evangelicals. Individualism, they argue, is part of evangelicals' "cultural tool kit," a mental repertoire that evangelicals apply when thinking about a variety of issues.[14] Not all cultural tool kits are religious in nature, but since faith is central for evangelicals, it infuses their outlook. Furthermore, although individualism is a broader American value, it is heightened within the evangelical tradition, being rooted in the theological sense that "man [is] a free actor . . . essentially unfettered by social circumstances, free to choose and thus free to effect his own salvation."[15] Historically rooted in the Reformation, this "accountable freewill individualism" has flourished among American evangelicals, becoming even more concentrated after shedding its progressive elements in the wake of the fundamentalist-modernist split.

As Emerson and Smith explain, because evangelicals emphasize personal relationships (flowing from their emphasis on having a personal relationship with Jesus Christ), they tend to believe that social problems are "rooted in poor relationships" rather than structural or systemic problems.[16] In the context of race relations, Emerson and Smith argued that this outlook not only rendered the structural problem of race invisible to evangelicals but encouraged them to view "blaming the system" as a wrongheaded attempt to shift blame away from the individual.[17] Similarly, my informants conceived of environmental behavior in very personal terms—the misbehavior of individuals—and actively criticized those who sought broader, structural solutions.[18]

THINKING LOCALLY

Scale has been an important issue for many environmentalists. When environmentalism first captured the public's interest in the 1960s, one of its key messages was that the planet was a finite resource and that growth could not continue indefinitely.[19] That the global scale matters to environmentalists is also implicit in many of the key issues about which environmentalists have raised the alarm over the past fifty years, including overpopulation, acid rain, ozone depletion, biodiversity loss, and climate change.[20] Titles of well-known publications such as the Worldwatch Institute's annual *State of the World* reports further illustrate the importance of global-scale thinking to environmentalists. Similarly, the New Ecological Paradigm scale includes two items that tap whether a respondent thinks about the environment's ability to support life in global terms.[21] The environmentalists I interviewed in Georgia also emphasized the importance of thinking on the global scale. Jim from the NEC argued that being unaware of the global scale of environmental problems was a fundamental impediment to recognizing the need for environmental activism:

> If people don't think anything is wrong—we're not running out of resources, we don't have overpopulation threats, we don't have worldwide competition for scarce resources, we don't have ocean acidification, we don't have air quality problems—you know, if a person believes that basically there are no problems at all, except that maybe they need cheaper gas, then that person is not going to be in favor of doing anything. Why should you do anything different, if everything you are doing is working fine, and there's no problem? *That's* the real problem.

As Jim pointed out, becoming an environmentalist entails understanding how local processes can scale up to create global disasters.

By contrast, my evangelical informants seemed concerned principally about local issues. This tendency to think on the local scale was underlined when they talked about their relationship with the environment in terms of how they themselves interacted with it, rather than talking about how humans in general interact with the environment. Often, when my informants emphasized the local scale, it was in order to exclude climate change as a matter of legitimate concern. Roger, pastor of the Independent Baptist church, told me, for example, that few people he knew took climate change seriously. Instead, they were more focused on local problems:

To the average person living in our counties, they're just concerned about whether or not the fishing and the hunting's affected because of what the [nearby] paper mill is doing. And if they're dumping something in the river right now that's going to affect or kills us, then we need to do something about it. But they . . . could not care less whether or not the polar ice caps are melting. They don't care. They just want [to know] they got fresh water to drink. It's a more practical view of life.

To Roger, it was outsiders—whom he elsewhere described as being on the liberal side of politics and looking down on "the average redneck in . . . Georgia"—who cared about climate change. The "global warming crowd," as he called those who were concerned about climate change, had their "heads in the clouds," whereas people in South Georgia were more practical—that is, more concerned about things that impacted them personally and on a local level.

Like Roger, Sheldon, a technician in his forties who attended a Church of Christ, argued that it was more important to take care of the local environment than to be concerned about global problems. As he told me, "we *can* affect climate, but we can't affect it to the point where the earth's going to cease. And I think ultimately, then, you have to go back to bringing it back to a smaller level, on a local level." In general, my informants preferred to think about behaviors that they themselves could control. Following Sheldon, many of them talked about not littering and recycling when discussing how to make sure they were being good stewards. Such comments mirror the individualistic ethos that Katharine Wilkinson observed in her focus groups among southern evangelicals.[22]

It is true that environmentalists have also emphasized the importance of acting locally, but this is usually in the context of awareness of the larger scale, as captured in the slogan "Think globally, act locally." While my informants did prefer to act locally, they did not tend to frame these actions in terms of their global impact.

The different ways in which the evangelical and environmentalist communities tended to think about environmental issues may even have been a reason why the NEC was struggling to build a local constituency. Though it had about five hundred local members, Jim lamented that "literally less than a dozen actually show up and do anything." Cathy, who also volunteered with NEC, had noticed that local environmental issues drew in a larger crowd than other ones that they were working on at a national level:

There are plenty of people I know that actually care about the environment around here. And they will work on other environmental issues like dolphins or protecting water quality or something like that. . . . But they are climate skeptics because they are conservatives. They are conservative Republicans, who care about their local river or something. And they want to protect it and they will fight a polluter, believe it or not. . . . But that's as far as they go.

Cathy's observation suggests that people from across the political spectrum were concerned about local environmental issues, yet this was not the case for national or global issues.

While the NEC was struggling, the situation was different at a locally focused environmental nonprofit that I call the Clean Water Coalition (CWC). Deborah, who headed the organization, told me that it had over 1,400 members, and that the number had spiked to over 2,200 on Facebook after a recent local environmental disaster (the organization's size was even more remarkable because it was based in a town less than half the size of the town the NEC was based in). The organization also had a positive, working relationship with local churches. When I asked if her group ever worked with them, she said that churches would sometimes ask them to give presentations about their work, and that they had worked closely with two Baptist churches in the region on separate issues—increasing energy efficiency at one church and helping to prevent a landfill from being sited near another.

One reason the energy efficiency collaboration was successful, according to Deborah, is that the CWC broached the issue of energy efficiency in the context of a campaign to prevent a coal-fired power plant from being located in the church's region. While the church's members were, by all accounts, unconcerned about the global problem of carbon dioxide in the air (a problem caused partly by the burning of coal), they *were* concerned about the local issue of mercury in the water, which threatened their ability to consume locally caught fish. Thus, by working on local issues, the CWC was able to develop a positive working relationship with the local community, including the local religious community. Meanwhile, the national environmental organization, which often focused on national and global issues, struggled to build an active membership.

The tendency to think on the local scale also came to light during an interview with Allison, who headed the Georgia chapter of a national religious environmental organization, which I will call Partners for Christian Stewardship (PCS). As Allison told me, the strategy her group had developed for talking with Christians about climate change was intentionally designed with this tendency in mind. When she or other members of the organization talked with churches about climate change, they never put it in a global context. Instead, they focused on its local effects. The tactic was blatantly pragmatic. As she put it, "what works is *localizing* conversation about climate change" (emphasis mine). When talking about climate change, they encouraged people to ask, "How is climate change affecting Georgia? How is it affecting our peanut crops? How is it affecting our rivers, our fish, the way we hunt? How is it affecting our air quality?" As she readily admitted, "We look very little at the rest of the country or the rest of the world. How is it affecting us right here? How can you look out your window and see how climate change affects things? And that affects people more."

Clearly, the tendency to think on the local scale that I encountered in the communities where I worked was something Allison had experienced in her work with Christians in other parts of Georgia. Notably, it is also apparent in her references to fishing and hunting that PCS had noticed that, to the extent that most people connected with environmental concern, it was through consumptive activities. The practical environmentalism that I observed in my focus groups therefore seemed to be a broader phenomenon.

GOD'S TOP CREATION

Christians have long been accused of being unconcerned about the environment because their tradition is anthropocentric, its ethical thought focused on human needs and not on the needs of other living beings or ecological assemblages.[23] Although some Christians have argued that the charge is misplaced (or does not preclude environmental concern), many of my informants proudly espoused anthropocentrism.

Pastor Philip, who headed an independent fundamentalist church in Dixon, agreed that it was important to be a good steward of creation but specified that he would prioritize humans over other species in doing so:

> Being good stewards of the creation? Great. I'm not [saying] let's throw garbage all over the street, but the value of human life is going to be more precious to me than the value of things, because, again, my theology says man is made in the image of God, so man is the highest priority. It's not that we burn the earth and save the man, but when it comes to choosing between the puppy or the person, I go with choosing the person, because they're made in the image of God and have eternal value. Jesus died for them.

While Philip gave lip service to stewardship in a practical vein, he quickly dismissed it as a topic for further conversation, proceeding immediately to a defense of anthropocentrism based on his reading of Christian theology. His views were exactly what Lynn White Jr. had raised the alarm about in 1967.[24]

Many of my focus group informants defended anthropocentrism actively, though the reasons they gave for supporting it varied. Ethan, a pilot in his thirties who attended a Presbyterian church near Mayfield, felt that humans should be given priority over animals because unlike animals, humans had souls. Meanwhile, Andrea from the Assembly of God church complained that protecting endangered species sometimes cost jobs. Claire at the Presbyterian church argued that "[God] created man to subdue the earth. So He wanted man to have dominion and rule, so I think He gave us the earth for our good." She defended this anthropocentric position by arguing that "if you're [taking care of the earth] for God's glory that's great, but if you're doing it for yourself or for the earth, I don't think that's what God intended His earth or you as a creature to be created for." For Claire, "doing it for the earth"—a passable definition of ecocentrism—would lead away from God's path. Upholding anthropocentrism was thus essential to her understanding of her faith.

I often heard statements about the need to take care of "what God has blessed us with," but these were often articulated within a framework of humanity's rightful dominion over the earth. There was virtually no debate among the people I met over whether human needs should take priority over those of earth's many other species.[25]

APPRECIATING NATURE

While my informants were only moderately concerned about the environment, to stop here would be misleading, for, especially in the rural areas, they were both knowledgeable and concerned about the health of the natural environment surrounding them. Often this was based in experiences hunting, fishing, or gardening. If there were any potential points of connections with the types of concerns that environmentalists harbor, it was here.

William, pastor of an Assembly of God church in Mayfield, spoke with particular feeling about his observations of environmental changes. Describing himself as an avid sportsman who loved to be outdoors, he apologized several times for his rambling comments on the topic.

During the course of our conversation, he both rhapsodized about spending time outdoors and lamented the changes to the local environment that he had witnessed during his lifetime. In the habit of going fishing on a weekly basis, he described how

> when I'm out there, I don't use the word *angers,* but it bothers me to see fishermen throw their bottles overboard, or cans overboard, or see them throw plastic overboard. . . . I guess I'm a little more passionate about that than others because I love to be outdoors. There's nothing to me more beautiful than going out in the boat headed east as the sun's coming up on the horizon of the ocean. It breaks your heart to get out in the ocean fifteen, twenty miles and see islands of debris floating by.

He went on to mention that much of the wildlife that had been present while he was growing up had declined. Water ducks and oysters had both declined, in his view possibly because developing coastal lands had disrupted their habitats. He also described the wildlife that he used to be able to observe when walking through the region's native longleaf pine forests. The pine forests used to "go on for miles and miles," but now they were diminished, and the fox squirrels, owls, indigo snakes, and whip-poor-wills that used to flourish there were now few and far between. It seemed clear that not only had Pastor William gained significant knowledge of his local environment as a hunter and fisherman, but he had noticed and was upset by the decline in species diversity and numbers that he had witnessed within his lifetime. He did not embrace the "environmentalist" label or any of its canons, but he was aware of and concerned about changes in the natural world. He cited this concern as a reason he had installed a recycling bin at the church (Mayfield did not offer home recycling services). Hence, while the majority of his and the church's community service efforts were directed toward human needs, such as providing food and shelter for those in need, the environment was evidently on his mind. To get ahead of my story a little bit, William was also a passionate end-time believer.

Another experience, this time outside the formal setting of an interview, underlined my rural informants' close connection to the natural world and resulting concern. On a summer evening in 2012, as my time in the field was drawing to a close, my husband and I were invited to a cookout with friends from the area. The dinner that evening was composed primarily of foods that people had hunted, fished, or grown. Although no one there would have described him or herself as an environmentalist, this was probably the most locally sourced meal I have ever consumed. Claiming a chair around a table piled high with dishes, I introduced myself to a middle-aged couple who were already seated and bantering with others. Dan and Sherry were longtime residents of the region, and from comments they made, I gleaned that they were Christians. When someone complimented Sherry on her squash casserole, for example, Dan good-naturedly protested that he should get credit for growing the squash. "God grew them, you harvested them," Sherry corrected crisply.

Given my interests, I told Dan about my project and asked if I could interview him. When we spoke later, he told me that he had grown up outside of Atlanta but had spent time in a rural area near Mayfield and, having family there, had returned as an adult. In terms of education, he had an associate degree in forestry and worked as a biological technician for a natural resource management agency. He did not identify as a Democrat or a Republican, but rather voted for whomever he thought would do a better job (as he put it). Based purely on this background, one would not expect Dan to be particularly concerned about the environment. Studies of the American public at large have found that the environmentally concerned tend to be well educated, politically liberal Democrats, raised and currently living in urban areas and employed outside of primary industries.[26] Dan had none of these characteristics, yet he clearly cared about the environment. He described how he had always liked to spend time outdoors and had chosen an occupation that would allow him to continue to do so. Additionally, a number of features of his lifestyle were environmentally friendly. In particular, he raised chickens, gardened (freezing and canning some of the produce), and hunted deer, which meant that his diet included many locally grown foods. Dan did not participate overtly in any environmental activism, but neither was he unconcerned about the natural world. Instead, he interacted with it on an everyday basis, and his concern for it was based on this interaction. At the same time, like most of the people around him, he went to church on Sundays (a conservative church of the Primitive Baptist denomination). Like Pastor William, he demonstrated how conservative Christian faith and environmental concern could blend.

For those of my informants who lived in rural areas, their everyday interactions with the outdoors clearly shaped how they related to it and what they valued about it. They read the environment not through the theoretical prism of global ecological degradation but through the prism of their own experience.

ULTIMATE IMPORTANCE

A final piece that is essential to understanding my informants' attitudes toward the environment—and, in particular, their emphasis on moderation—has to do with their faith. My informants' sense that what they currently did was sufficient might seem simply defensive ("Don't tell me what to do") if it were not so clearly linked to a larger sense of order that was rooted in religion. As they repeatedly told me, the environment was important, but it was not *ultimately* important. Hence, to spend excessive time worrying about environmental problems was pointless and wasteful, a distraction from the real purpose of life. One way to interpret this is to see it as confirming my informants' actual lack of environmental concern. But because such sentiments were so often expressed in the context of moderate concern for the environment (as discussed above), I find it more useful to think about such statements as reflecting my informants' desire to integrate the environment into a framework that was meaningful for them. Notions of both salvation and sin seemed to be particularly salient in this regard.

My informants frequently noted that the environment was not ultimately important because salvation should be one's earthly priority. For Michelle, the teacher from the fundamentalist Baptist church, for example,

> when I see people so concerned about climate change—and, not just climate change, but you know, saving the animals, and saving the sharks and the whales and the chipmunks—and I see the concern that they have that their world is falling apart, it makes me extremely concerned for their souls. . . . I see we have so many people around us that don't have the hope that we have. Because I think—and I'm going to be honest—to me it's trivial to be concerned about those things. . . . Because I think the real issue is reaching their souls. Because guess what, this isn't going to be forever. Heaven's forever.

In Michelle's view, people who were worried about climate change and other environmental problems had their priorities askew. If when you die "you either go up, or you go down" (as a concerned informant reminded me) and wherever you go you stay for eternity, devoting your time on earth to protecting the environment is a trivial pursuit indeed.

This logic did not just move environmental stewardship down the list of priorities, it evacuated such activities of deeper meaning. As Sheldon at the Presbyterian church put it, "environmentally speaking, yes, Christians should be the strongest advocates for making the world a better place. . . . But ultimately does it matter? As long as we teach Christ crucified and everything that comes with it, it don't matter." Why did my informants so often affirm that they were concerned about the environment in the same breath that they denied the significance of such views? Typically, as these examples illustrate, they did so not to endorse wasteful behavior, but to underline the centrality of faith.

A second common way in which my informants reinterpreted the meaning of the environment—and, in particular, environmental degradation— in light of their faith was in the context of sin. I first noticed this type of reinterpretation during a focus group I conducted at the same Independent Baptist church Michelle attended. I had sensed restlessness throughout the discussion, but unable to pinpoint its source, I pushed on. Toward the end of the discussion, Kenneth, a fifty-something technician for a local pulp mill, ventured that I was missing some important background information. In order to understand their answers, he said, I first had to understand that

> there is a spiritual context to this. . . . When sin entered the world [with Adam and Eve], death entered the world and corruption entered the world. And our world—I know this is going to sound *crazy* to somebody who's not a Christian—our world is being corrupted through storms and earthquakes and pollution *because* of mankind's sins. That's where *I* have a part to play. I believe that the Bible teaches clearly that our sin, and I'm talking about mankind now, is corrupting this world, and that's why the world is going down, you know, it's not as pure and as good as it was. And it's going to continue to go down until Jesus comes back. (emphasis his)

My first instinct was to try to make sense of his comments in terms of my own worldview, so I asked hesitantly whether he meant that sin caused environmental pollution. But he rejected this interpretation impatiently, saying, "I'm not talking now about things like polluting the world. . . . I'm speaking about just sin . . . you know—rebellion against God, that's what sin is." His dramatic phrasing made me pause. Acknowledging the beat, he caught my eye and added, "I know, that's a sobering thought, isn't it?" It certainly was.

Reflecting on the exchange later, I realized that his statements had accomplished three things. First, he had reframed environmental problems in religious terms, so that instead of them being problems in and of themselves, they were presented as the result of a deeper religious problem (sin) that could only be solved through religious means (salvation). He then secured this reframing by thwarting my attempt to bring the conversation back to the environment, saying that the environment was not what he was talking about. His comment also underlined the point that pollution itself was not important; it was the *cause* of pollution that was important. Finally, he made the story powerful by making it personal: he wasn't just talking about sin in general, but about *my* sin. Ultimately, what he was attempting to impress upon me was that my previous line of questioning was misguided—I was asking about things that were insignificant, and ignoring the things that were of utmost importance. Environmental degradation as a problem in and of itself did not connect meaningfully to the group's understanding of the way the world worked. But environmental degradation did make sense in light of sin.

After this encounter I paid closer attention to my informants' references to sin, and I began to see that they understood sin as the causal force behind much of what was negative in the world. Philip, the fundamentalist pastor in Dixon quoted above, did the most to help me understand this perspective. A youthful and energetic graduate of Dallas Theological Seminary, he had a quick wit and an even quicker tongue; I scrambled to keep up with my notes as we talked. When I asked him whether he thought Christians had a particular obligation to care for creation, he agreed, but then added that some things, like volcanoes, could not be stopped. A murmured "of course," was on the tip of my tongue, but before I could respond, he added that the reason they could not be stopped was because they were the result of sin. Surprised, I asked what he meant. He explained that in the beginning God had created a perfect world, but when sin entered the world it brought along with it certain negative consequences:

> All of creation fell when Adam fell, and so things went into motion there that initially weren't there. So now in creation there are struggles and thorns and storms that happen as a result of sin. And so why do bad things happen to good people? Ultimately it's not because God is just mean, it's because it's the result of sin. Because sin entered the world there's now consequences for that and unfortunately sometimes, earthquakes and these things happen. That wasn't the original design, but that's why there's the need for a savior and a redemption, and a return biblically of Christ to redeem and to restore.

From this perspective, volcanoes and earthquakes existed because creation was still suffering the consequences of Adam and Eve's original sin of disobeying God. Imperfections in the natural world, including imbalances in the weather, were thus the result of sin—original sin. Significantly, diagnosing the problem in this manner allowed him to argue that his worldview offered what was needed to solve it: a savior and redemption.

The idea that sin causes environmental degradation is hard to reconcile with the idea that humans cause environmental degradation. These are not just two different perspectives on a problem, but two different notions of causality and visions of reality. In the version of the story that my informants presented, the root problem was sin, and the solution was salvation. Environmental problems were window dressing in this story, showing that creation was cursed, but not having intrinsic significance. Even if environmental problems were substantial, the best way to address them would be to tackle the root of the problem: sin. And this was exactly what my informants often suggested. Cam, from a Lutheran church in Dixon, argued for example that "what you see wrong in the way humanity uses creation is the direct product of sin that we did. And if we were not sinful people—and I'm including every person that takes breath—if we were not sinful people there would be no issues about the environment."[27] Yet what is interesting here is that even though Cam attributed environmental misbehavior to sin, he (like nearly all my informants) acknowledged that humans abused their environment and that doing so was wrong. To dismiss such a reinterpretation as basically anti-environmental requires ignoring an important point of agreement.

This gets at one of the key dimensions of the conflict between evangelicals and environmentalists. While my informants saw the kinds of sentiments expressed above as compatible with moderate environmental concern, many environmentalists would view them as cringe-worthy proof of what environmentalists have suspected all along about evangelicals: that their home is elsewhere, and so they do not care much about the fate of the earth. Curiously, however, both groups approve of many of the same behaviors (replanting trees, recycling, not polluting, and the like). This suggests that the problem is not always the types of behaviors that the groups endorse, but the different worldviews that give those behaviors meaning. It is as if the evangelical and environmental worldviews held opposite polarities, orienting them toward heaven and toward the earth, respectively. Despite their similar composition, when one tried to bring them together, they repelled each other.

Talking with my informants at Pinewood Baptist Church further demonstrated how this polarity operated. Alyssa and Justin were a young couple in their twenties whom Jill had befriended while they were in high school. At the time I met them, Alyssa was working in a local elementary school in a small town while Justin was employed as a corrections officer at a nearby prison. Both were intelligent and articulate; it was also evident that Jill had encouraged them to think about the environment, so they had had the chance before I met them to wrestle with how to combine their faith with concern for the environment. Both Alyssa and Justin repeatedly argued that Christians should be environmentalists. "I feel that environmentalists can overlap with Christianity," Alyssa wrote me in a note after the focus group at her church. "The knowledge that we gain from the environmentalists is useful for Christians." Thus, in contrast to many of the other Christians I had met (and clearly thanks to Jill), Alyssa and Justin had a positive view of environmentalism; they were aware of common negative stereotypes about it but generally chose to ignore them. Still, their familiarity with how most people around them regarded environmentalism gave them a uniquely insightful perspective when it came to barriers.

Once, when I asked Justin about whether Christians and environmentalists might work together (citing a specific example of signing a petition that had circulated recently), he expressed doubt and suggested candidly that part of the challenge was a desire not to disrupt the status quo: "We [Christians] are a very scared people, who if we stand and do anything, [worry that] people may look at us or talk about us funny." When I encouraged him to expand, he added that despite his sense of common cause with environmentalists (in the abstract), in reality working together would be difficult because "not all, but some of the environmentalists, they believe in more of a—I may be wrong—but most of them believe more in a, you know when you die, you are going to go right back into the earth." Acknowledging that this idea was probably to some extent a stereotype, he nevertheless added that it posed a real barrier:

> It's really hard to work with people that basically are almost opposed to what your belief system is. . . . People who don't believe in Jehovah God and salvation would have a hard time being with the people that that's all they talk about and that's what they believe in.

As Justin underlined, the barriers to cooperation were fundamentally social: not just about conflicting ideas but about the difficulty of getting along with people who look at the world differently. To put it another way, collaboration requires finding not only those with whom you can agree, but those with whom you can feel comfortable, who operate within your same magnetic field.

Perhaps this explains why moving between fields gave me a kind of double vision. There was a moment after a few months in the field, for example, when I was surprised to find myself annoyed by a joke about James Dobson on *The Daily Show,* which seemed to be an obnoxious mischaracterization of his views. The double vision also gave me a heightened sense of irony. One day when passing by the marquee of a little countryside church, I noticed the following message: "Drought got you down? Come bathe in the living waters of Christ." Viewed through an environmental lens, it was the perfect expression of Christian environmental indifference. But viewed through a Christian lens, it spoke powerfully of the restorative potential of faith in the face of adversity.

DEFIANT MODERATION

The political scientist John Dryzek's scheme for classifying environmental discourses provides a helpful map for summarizing how my evangelical informants' sensibilities compare with other environmental discourses. Dryzek analyzed the various environmental discourses that have evolved since the inception of the environmental movement by assigning each a position along two axes: reformist-radical and imaginative-prosaic. The reformist-radical axis tapped the degree of change that was sought—minimal or drastic. The imaginative-prosaic axis tapped how change was envisioned—as fundamentally restructuring versus tinkering with the current industrial-capitalist system.[28] In these terms, my informants' viewpoints were both reformist and prosaic. They thought any change should be minimal, basically amounting to broadening the sphere in which "good housekeeping" was practiced. And they were prosaic in the sense that they took the industrial-capitalist system as a given.

But my informants' environmental attitudes were distinct not just in their content, but in their tone. A Southern Baptist pastor with whom I spoke toward the end of my fieldwork, Mike, captured this basic ethos perfectly. His response particularly surprised me because he had signed a declaration that supposedly expressed support for taking action on climate change (discussed in chapter 6). But he was no environmentalist, at least not in the way that term is usually understood. When I asked him if he could ever see himself getting involved further in environmental issues or whether it was lower on his list of priorities, he rejected my framing of the question impatiently, responding with an air of defiant moderation:

Well, it's neither one of those. It's C, none of the above. Because I wouldn't say that it's low down on my list of priorities, but I also wouldn't see myself being more involved. I think I'm sufficiently involved. My family, specifically, we are very careful to recycle. We are very careful to take care that we don't waste. I'm an outdoorsman. I'm very careful what I do outdoors. If it's hunting or fishing, I do it in a responsible, reasonable way. I also watch out for what I'm doing—not to just leave a trail behind me of trash and things like that. So we are conscientious in that respect. Do I see it getting above and beyond that and encouraging others to do the same kind of thing? No. But I also don't consider that low on my priorities either. That is just something we conscientiously do.

Many of the other Christians I spoke with in Georgia expressed a similar sense that what they currently did was sufficient. Common sense dictated cleaning up after oneself, but social propriety dictated that such behaviors be accomplished without fanfare or disruptive behavior.

This desire to avoid conflict points to what may be a major reason why my informants, and perhaps Southern evangelicals more broadly, have not embraced environmentalism more enthusiastically. In many ways it went against the grain of their religious and cultural commitments. Asking someone to alter either, not to mention both, is asking a lot. It requires a reordering of perceptions, values, identity, and the social relationships that sustain them. As my conversation with Justin and Alyssa showed, it was possible to do, but not easy.

Belying many environmentalists' hopes in the mid-2000s that a potent religiously inspired form of environmentalism might be emerging, my experience among laypeople in Georgia suggested that faith often stood in the way of environmental activism, while paradoxically being compatible with locally scaled, individualistic, and private pro-environmental attitudes and actions. In the next two chapters, I explore how end-time beliefs were connected to these practical, defiantly moderate environmental attitudes.

CHAPTER 3

End-Time Beliefs and
Climate Change

On an evening near the end of February 2012, I was sitting in the octagonal great room of a Pentecostal church in the suburbs of Dixon. I had arrived at a Wednesday evening "life group" expecting to do a focus group, but the room was chaotic and, because of a miscommunication, no one had come expecting to stay for an extra hour to speak with me. Seeing no reason to give up on the evening entirely, I sat down to chat with Reid, who was from the region, and his wife, Paulina, a friendly woman originally from South America. When I asked them about how their faith might relate to the environment, they—like several others with whom I spoke that evening—had to think for a while before finding a connection, which suggests that it was not a highly salient issue for them. But when I shifted the topic to ask what they had heard about climate change, Paulina did not hesitate to answer matter-of-factly that "all [that] is happening right now, is what the Bible said in Revelation. [It] has been happening because it's supposed to happen. It's the end, almost the end. I think that God is close to coming and that's why all this is happening."

With these words, Paulina qualified as what I will call a *hot millennialist,* or someone who believes that the end is near and readily views climate change as evidence of this fact.[1] Only a handful of my focus group participants fell into this category, but their views matched the portrait that critics have painted of end-time believers remarkably well. Not only did they interpret climate change as a sign of the end times, but they were excited by the idea that the end was near—although they were also saddened that they would be leaving unsaved loved ones behind. Just as critics have suggested, they seemed unconcerned about the fate of a planet they believed they were destined to leave.

But hot millennialists like Paulina were in the minority. The majority held a different view, namely, that the timing of the end was not imminent, but unknowable. Importantly, this group's views about climate change differed as well. Arguing that climate change did not fit into the prophetic scenario, *cool millennialists* were predominantly climate skeptics.

That I encountered two types of end-time believers reveals something important about end-time believers in general. Again, most discussions about the environmental impacts of end-time beliefs focus on the attitudes of premillennialists, among whom belief in the imminent end of the world is thought to predominate. However, my time in the field amply illustrated the dangers of assuming that theological categories match social reality. In particular, even though my focus groups were divided between premillennial and amillennial churches, the end times came up without any prompting in eight of the nine focus groups. Premillennialists were clearly not the only ones who were thinking about the end times. In fact, the major division I found among end-time believers is not based on how their churches or denominations interpret the millennium. Rather, the major division occurs at the individual level, between those for whom the end times are a highly salient, top-of-mind issue and those who believe, as a matter of doctrine, that Jesus will return to earth someday but who emphasize that, as stated in scripture (Matt. 25:13), they can "know neither the day nor the hour."

It is important to emphasize that hot and cool millennialism are not static categories. My interactions with churchgoers in Georgia made it clear that interest in the end times can wax and wane over the course of a lifetime (and may never take root at all). But at any given moment, people can generally be classified as falling into one camp or the other—and these distinctions are worth making since they corresponded with different attitudes toward climate change.

HOT MILLENNIALISTS, THE END TIMES, AND CLIMATE CHANGE

There was at least one hot millennialist present in four of my nine focus groups, but in only two of the groups—those held at the Assembly of God and Seventh-day Adventist churches—did participants discuss these views extensively.[2] I doubt this is a coincidence, given that eschatology has long been emphasized by both Pentecostals and Seventh-day Adventists; end-time belief is more mainstream in these churches, which presumably makes people feel more comfortable expressing such views (and also increases the likelihood that hot millennialists would be members of those particular focus groups). The hot millennialist views expressed at the Seventh-day Adventist and Assembly of God churches did not differ substantively from what I heard at the other two churches, so to simplify the analysis, I will draw on those comments made in the Assembly of God and Seventh-day Adventist focus groups.

As a Pentecostal denomination, the Assemblies of God accepts the doctrine of premillennialism. In fact, premillennialism was a foundational aspect of the Pentecostal theology that Charles Fox Parham, the movement's founder, developed in the early 1900s. The "gift of tongues" is a defining feature of Pentecostalism and is interpreted as an eschatological sign of the return of the Holy Spirit.[3] Over a century later, the end times continue to be an important denominational focus. In 2013 the Assemblies of God's official website cautioned that one should not read biblical prophecy as one reads a horoscope, but bluntly added that "the Assemblies of God preaches a clear message that Jesus is coming soon, and we will preach that message without apology, no matter how long the Lord delays His coming."[4] A separate page on "The Second Coming," which is listed as a core doctrine, interprets current events according to biblical prophecy: "With the world experiencing natural disasters, economic downturns and increasing uncertainties on many fronts, the doctrine of Christ's Second Coming is more relevant than ever."[5]

Seventh-day Adventists also place great emphasis on speculation about the end times. In fact, the denomination grew out of the remnants of an apocalyptic movement led by a farmer turned preacher named William Miller, who predicted Jesus would return to earth sometime between 1843 and 1844. When his prediction failed, many left the movement, but others continued on, arguing that important prophecies had been fulfilled on the day Miller expected, and that Christ's return would soon follow. Led by the prophetess Ellen G. White, Seventh-day Adventists (as they became known) continued to expect Jesus to return soon. When he did not, the denomination began to put down firmer roots in society—ostensibly in order to spread the message that the end was near—opening hospitals and schools and sending missionaries abroad. As the sect developed into a more established denomination, Adventists became divided on the issue of eschatology. It remained the centerpiece of Adventist evangelism, but according to a study by the sociologist Ronald Lawson, by the 1990s only a small minority of pastors emphasized it in established churches. When pastors did mention eschatology, they typically discussed it as a matter of doctrine, rather than attempting to instill a sense of urgency.[6] Seminary teachers also downplayed its importance.[7] Thus, Seventh-day Adventism, while historically nurturing apocalyptic sentiments, has become divided with regard to eschatology.

My analysis of the focus groups revealed a number of commonalities in the ways that hot millennialists at the Assembly of God and Seventh-day Adventist churches talked about climate change. Most importantly, when the topic of climate change came up, they immediately and confidently linked it to the end times. As environmentalists feared, this interpretive framework rendered climate change meaningless in and of itself; it was a sign that end-time prophecies were being fulfilled, rather than that humans were destroying the environment.

Climate Change as a Sign of the End Times

In both the Seventh-day Adventist and the Assembly of God focus groups, the topic of global warming seemed to immediately call to mind end-time prophecies. Tied to this was the notion that if global warming was a sign of the nearing end times, there was nothing anyone could do to stop it. In the Assembly of God focus group, I had asked the group whether they considered harming the environment a sin. Sarah, a young mother, replied that she did think purposefully harming the environment would constitute a sin. Craig, a member of the armed forces in his thirties, responded by cautioning that it was important to "draw the line between protecting and worshipping" the earth. The conversation then veered sharply (from my perspective), as his wife Julie spontaneously brought up the topic of global warming:

Julie: Because, like, with the polar bears and stuff, you know, of course I don't want them to die, but you also have to realize this is just a part of the world coming to an end like it's supposed to. And there's nothing really that they can do.

Sarah: Yeah, we can't stop it.

Julie: That's why we need to be educated in the Bible so we know what signs to look for. Because you're just wasting all that money on research when it's, sadly, not going to help.

Robin: Would you say more about that? If anyone else wants to jump in. Like, why wouldn't it help?

Julie: Because that's just how God planned it. I mean it's just going to keep—
the ice, it's going to keep dissolving, the temperature is going be different.

Craig: It's going to be hotter—

Julie: —from where it was before. Because if you just—like I've noticed over
the past three or four years on the Weather Channel that every day there's
a record: "This hasn't happened in a hundred years, this hasn't happened
in fifty years." And there's a reason for that, it's because Jesus is coming.

This spontaneous discussion of climate change as the fulfillment of end-
time prophecies leaves little doubt that these individuals viewed it as a sign
of the end times. Sarah and Julie also both emphasized the futility of trying
to stop climate change, saying that "we can't stop it" and that conducting
research would not help because "that's just how God planned it."

Hot millennialist participants at the Seventh-day Adventist church
also seemed to view global warming as the possible fulfillment of end-
time prophecies, although here there was a greater diversity of opinions.
When I introduced the topic, James, a college student, responded that
he found the notion of anthropogenic climate change to be credible
based on his own observation of climatic changes. He added that "I
think it's not an issue that we should be ignoring." Kirstin, a middle-
aged daughter of Haitian immigrants, shifted the topic to prophecy:

Kirsten: To some extent, I don't know if everybody will agree, but I think
that the global warming issue, because like you know the pastor mentioned today, that terrible things will happen in the end of days, and I
think that to some extent that's part of maybe what we may not have
total control over. Because I do believe that no matter what, there will be
those disasters. In fact, the Bible said those things will happen.

Leila: [Agreeing] Mm-hmm, will come to pass.

Kirsten: Yeah. And yes, we may be able to do better with our gas. Like,
there's things that we may be able to control, but in the end I think it's all
going to have to come to pass for the Second Coming of Christ. It's just
what the Bible says. And I believe it, and those things, we don't really
have much control over it.

James: I think that's actually one of the plagues. In Revelation, it says the
sun will scorch man, and they're going to curse God. So obviously something has to happen for that to happen. It's already getting there.

Similar to the Assembly of God focus group, Adventist participants, particularly Kirsten, immediately linked global warming to end-time prophecy. Although James had already indicated that he thought global warming should be addressed, he evidently also considered it possible that it

fit into the prophetic scenario. Kirsten also emphasized the futility of trying to stop climate change by asserting that we "don't really" have much control over it. Importantly, however, by saying that "I don't know if everybody will agree," she acknowledged that others might not share her interpretation; and, in fact, opinions in the group were divided.

Back at the Assembly of God focus group, Julie and Craig emphasized the futility of human intervention again when I used the discussion quoted above as an opening to ask the group what they had heard about global warming or climate change. A few participants mentioned that it made them think of Al Gore, but when I shifted the conversation slightly to ask how they felt about climate change, Craig responded:

> *Craig*: Anything dealing with the environment is, that I mean to me is destined to happen. Everything that's happened is destined to happen, is supposed to happen.
>
> *Julie*: God allowed it.
>
> *Craig*: Yeah, God allowed it to happen. So to say that we can change that is, I mean, people can say all they want, that's because usually the ones that'll say stuff like that are not spiritually minded, because like [Julie] said, if they would actually read Revelation they would see what Revelation says about the earthquakes we're having, about the hurricanes. Everything that's happening is in Revelation. So instead of crying because we're not taking care of the earth, they should be happy knowing that the end is coming soon.

In Craig's view, trying to stop climate change was futile, because whatever was happening was God's will. At the same time, by criticizing those who were "not spiritually minded," Craig implied that those who were trying to figure out what to do about climate change were misguided and perhaps even spiritually dangerous.

To make sure I was understanding Julie and Craig correctly, I asked whether the changes they were seeing, including global warming, were anything to be concerned about.

> *Julie*: To me it's just prophecy.
>
> *Craig*: Yeah.
>
> *Julie*: I don't really get into the whole, this person said, and that person said, either. It's just easier for me to pray about it and let the Lord speak to my spirit, like, what is true, and just wait and see what happens.

Their responses left little room for doubt that they did not think anything could or should be done to address climate change.

Climate Change as a Symptom of Other Problems

A second commonality in the way hot millennialists viewed climate change was that they considered it (along with other environmental disturbances, such as earthquakes) to be symptomatic of other problems, rather than a problem in and of itself. As discussed in chapter 2, this sometimes meant they downplayed environmental problems in the context of larger questions about salvation and sin, but in other cases the end times served the same function.

At the Assembly of God focus group, when I asked participants if they thought humans were responsible for climate change, Craig responded that "it's Revelation coming true." He went on to criticize the effort that people put into analyzing historical weather records; for him, to do so was to allow one's attention to stray from what was more important: preparing oneself for Jesus's return, as described in Revelation.

When I asked the Seventh-day Adventist group whether they thought humans were responsible for climate change, Kirsten also made comments suggesting she did not find environmental decline troubling per se:

> I think something has to happen to make those Bible verses come true. So if not global warming then what? Something has to happen for, you know, all these things to happen before Christ can come. And maybe we do contribute. I mean, we're sinners like you guys mentioned before, and we do things all the time to damage ourselves, to damage our relationship with God, with men and with our environment. But . . . I really am not sure that our contribution is really, you know, that much bigger of a factor.

It was not so much that Kirsten didn't think humans were responsible but that she was uninterested in who was responsible. What she was more interested in was signs that biblical prophecies were being fulfilled. She weighed the possibility that humans were responsible (using the framework of sin), but ultimately she went back to the idea that climate change was a sign of the end times. In this context, climate change was not itself a problem, but rather a portent of things to come.

COOL MILLENNIALISTS, THE END TIMES, AND CLIMATE CHANGE

While hot millennialists accepted that climate change was occurring and viewed it as a sign of the end times, cool millennialists—who were the majority of my informants—were skeptical about climate change, emphasized that the timing of the end was unknowable, and doubted that biblical prophecies referred to climate change. They also differed from hot millennialists in explicitly arguing that end-time belief did not authorize environmental irresponsibility, although this did not make them enthusiastic environmentalists.

"Ye Know neither the Day nor the Hour"

Rather than believing the end of the world was imminent as hot millennialists did, my cool millennialist informants explicitly balanced their faith in the Bible's teachings about the end times with the need to live in this world. While they might refer to possible signs that the end was near, they would typically hasten to add that the timing was unknowable. They therefore tended to place far less emphasis on the end times than hot millennialists, and expressed less excitement about and investment in the prospect.

Sheldon from the Church of Christ exemplified this attitude. When I asked the group whether they saw signs that the end was near, he argued that knowing the end times could come at any moment did not necessarily change his behavior, except insofar as it encouraged him to continue to act righteously:

> The end of times is going to come when it comes. We're supposed to live every day like it's our last day. When I get up in the morning I need to make sure I'm right with God, when I go to bed I need to make sure I'm right with God. . . . Do I think all these things [current events that might be signs of the end] point to the end of times? I think it's all just things that happen. I don't know what it's going to be, I don't worry about it. It's not a concern of mine.

For Sheldon, as long as he was "right with God," the exact timing of the end was irrelevant.

The same seemed to be the case for Ben, the head pastor at an independent Church of God (a Pentecostal denomination). Ben was a premillennialist who speculated that the country's "moral freefall" might be part of end-time events. Yet when I asked him how knowing that the end times could be right around the corner affected his everyday life, he responded that it did not change much:

> Jesus said, "Occupy till I come." No man knows the day or the hour. We live like He's coming today, but we plan like He's coming tomorrow. So He said occupy, so that means to be busy, to maintain, to live victoriously. So I believe we've got to set goals and plans. We're looking at our . . . budgets and calendars, so I'm planning like I'll be here next year.

Being a premillennialist did not prevent Pastor Ben from making long-term plans.

Sheldon's and Ben's comments exemplify how aware my cool millennialist informants were of biblical teachings about the end times. Indeed, people in this category would surely have been counted as end-time believers in surveys that asked, for example, whether the prophecies in the Book of Revelation were accurate, or whether they believed Jesus would return to earth someday. Yet this did not mean, as the proponents of the end-time apathy hypothesis often suggest, that they welcomed climate change as a sign of the end times.

In comparison to hot millennialists, my informants were quite circumspect when it came to identifying current events as signs of the end times. When I asked the participants in the Southern Baptist focus group whether they thought the Book of Revelation applied to current events, they cautioned that it was difficult to know where the current moment fit in the prophetic timetable.

> *Graham*: It depends on what you're applying it to. If you're applying it to the environment, to some degree, maybe. Are we seeing more climatic catastrophes than we did years ago? Yeah. Can that be applied to Book of Revelation? To some degree, probably. You know when He says that there will be more natural disasters in diverse places, we've got things occurring in areas that never had what we had. We had one earthquake in Virginia on the east coast that people said, well we've never had that before. So yes, there's a possibility there. We don't know everything about the Book of Revelation like we like to think we do. As far as humanity, does the Book of Revelation apply to that? Yeah. We're seeing more things similar to Sodom and Gomorrah than we ever had. Society's changing in its behavior. And the Book of Revelation tells us that towards the end times that that's going to occur. You know. So does the Book of Revelation apply, to some degree yes. Is it the end?
>
> *Renee*: Nobody knows.
>
> *Graham*: We don't know. We don't know God's time. What's the end? Fifty years? A hundred years? A thousand years? We're not God.
>
> *Tom*: That's the issue with Revelation, is that we don't know what God's time frame is. You don't know how many centuries . . . I mean how long will all this play out. . . . Jesus was very clear that we won't know the day or the time or the hour when He will come back.
>
> *Renee*: *He* didn't even know. (emphasis hers)

Graham's and Renee's comments in particular illustrate how cautious my cool millennialist informants were about applying biblical prophecy to the present day. Most of these informants believed that things were getting worse, and that some aspects of prophecy were perhaps being fulfilled. Like hot millennialists, that is, they were comfortable suggesting that some of the negative changes they were seeing in society (especially the increasing societal acceptance of gay marriage, abortion, and profanity) could have to do with the end times. But unlike hot millennialists, they never fully committed to the belief that the end times were imminent.

Climate Change: Not a Sign of the End Times

From a biblical standpoint, it would seem relatively easy to argue that climate change is a sign of the end times. The verse in 2 Peter that links the end of the world with heat would seem to provide particularly strong support: "Looking for and hasting unto the coming of the day of God, wherein the heavens being on fire shall be dissolved, and the elements shall melt with fervent heat."[8] During the Cold War, many prophecy writers cited this passage in support of their argument that the Bible had predicted that the world would end in a nuclear war. But there is a long history of adapting prophetic scenarios to fit current events, so it would not be surprising to hear new claims about climate change's relevance to the end times.[9] Indeed, Jehovah's Witnesses (a non-evangelical denomination known for emphasizing end-time beliefs) have interpreted climate change in exactly this manner.[10] Yet the cool millennialists in my focus groups resisted this interpretation, further underlining the differences between them and hot millennialists.

For example, when I asked the group at the Independent Baptist church whether they saw any signs that Jesus would return soon, Michelle noted that

> in Second Timothy it talks about [how] there are lots of things that define what the last times are, it's not just having to do with the climate. The climate is like a *small* thing compared to what—I mean, the Lord doesn't list that in the top things. Mothers with unnatural affection, fornicators, liars, blasphemers, people that are proud, that kind of stuff is what He lists. He doesn't say, "It might get hotter outside." (emphasis hers)

She rejected the notion that climate change was a sign of the end times, citing lack of biblical support. Similarly, Kelly, a woman in her forties who participated in a focus group at a nondenominational Christian Church, rejected the notion that the climate was changing, adding, "I know how God says the earth is going to end, and I don't sit around fearing that the polar ice caps are melting and this and that." For her, climate change could not be happening because that was not how the Bible portrayed the end of the world.

In sum, while cool millennialists accepted earthquakes and natural disasters as indicators of the approaching end times, they found little biblical support for the idea that climate change could be interpreted similarly. Instead, as chapters 4 and 5 explore in greater detail, they tended to argue that those who suggested the climate was changing were doing so in order to *undermine* Christian understandings of the end times. Hence they tended to see climate change as a competing eschatology, rather than as fulfilling biblical prophecies about the end times. Clearly, cool millennialists' interpretation of climate change differed greatly from that of hot millennialists, who were eager to believe that it was a harbinger of the end.

Arguments against Environmental Apathy

While my cool millennialist informants believed that Jesus would at some point return to earth, they emphasized that this was not a license for environmental irresponsibility. They instead argued that their obligation to "occupy" until Jesus returned applied to the environment as well. Crucially, while such sentiments distanced them from hot millennialists, the notions of environmental responsibility they expressed were limited in all the ways discussed in chapter 2. Still, their rejection of end-time-inspired apathy undermines the claim that their environmental attitudes are a direct result of believing the end is near.

When I asked Richard, the pastor of a large Southern Baptist church in Dixon, whether believing in the end times relieved Christians of their responsibility to care for creation, he disagreed emphatically:

> No, He says be faithful unto death. We're to be faithful until He comes. The whole idea of the Christian faith is don't relax, don't let up, don't give up. . . . He says good is going to win, so make sure you're on the winning side. So that makes me active today. So we're not fatalists as far as the earth is concerned, [thinking] "Oh it's going to get burned up anyway, so I just can dump oil in the river." That's irresponsibility.

At the Church of Christ, several informants brought up the same point. Tyler, an army employee who was in his forties, had suggested that it was impossible to live without harming nature, since humans had to live somewhere, and doing so would always destroy habitat that other species might need. Sheldon pointed out, however, that this was not a license for carelessness:

> I think a danger that happens . . . is just in the church and in society in general [that] we'll become complacent about things . . . because we trust in God that He's going to make everything OK, but then we're also supposed to be doing our part to make it happen. So with nature it'll be the same thing. If as Christians we just say that God's just going to take care of it and it's all going to be fine because we're Christians . . . then it's going to start getting worse and worse and worse, and we're not leaving anything to our children and not taking care of that. And we have an obligation to make sure that our kids have the things that they need as well.

Sheldon suggested that Christians had a responsibility to care for the earth regardless of whether the end might be near, which paralleled my informants' general assertion that believing Jesus would return to earth someday did not change how they lived their lives (except to inspire them to remain strong in their faith). The cool millennialists to whom I posed this question unanimously shared this sentiment.

Tensions between Hot and Cool Millennialists

Underlining that hot and cool millennialists really were two separate groups, I found that their differing opinions about the reality of climate change put them at odds with each other. The result was that even though neither group thought anything should be done to halt or mitigate climate change, the two groups were not, with regard to this issue, unqualified allies. A series of exchanges among three participants in the Assembly of God focus group demonstrates how hot and cool millennialist viewpoints could conflict. Andrea was highly skeptical that climate change was occurring. As described above, Julie and Craig did believe that the climate was changing, and they saw it as part of the end-time scenario. The following discussion shows how this put Andrea at odds with Julie and Craig.

> *Andrea*: I'm amazed when I hear people in our church talk about the fact that the climate's changing, and global warming and all that. Because I really don't quite believe it. . . . I listen to Fox News a lot, and they don't believe Al Gore's science is correct, and they found where they had doctored some studies. And I might just have a lot of doubts about global warming even existing. I think it's self serving. I think there's a— [interrupted]

Julie: I don't know, but doesn't the global warming, does that affect the ice that is melting away? And a lot of the polar bears and seals and all those different animals—

Andrea: Well yes, but how do we know that one hundred years ago or five hundred years that didn't happen for a while and they'd go back.

Julie: Because I like talking to older people to hear their wisdom, and everybody I've talked to about the weather, [they say it was] nothing, nothing like it is now.

Andrea: But like you said one hundred years ago. I just, you know. We're on this earth for a short time to see just a little part of what happens.

Julie: I would say, I believe there is global—[both Andrea and Craig try to cut in at the same time].

Andrea: There may be. I'm not a scientist to know for myself, but I know that scientists disagree about it. And I tend to be for the other side.

Craig: Yeah but, I mean, I'm no scientist, but I know if it was hotter this time this year than it was this time last year, then.

Julie: It's just getting worse.

Andrea: And that's a short-term climate change, and that's not—you know, there's local weather. And there's global weather. But it all has to be taken in perspective of thousands of thousands of years, not just this year versus last year.[11]

Andrea was clearly well aware of skeptical arguments, an unsurprising fact given that she said she watched Fox News, which has presented skeptical arguments about climate change far more often than other news channels.[12] In terms of how researchers categorize climate change attitudes, her strong defense of skeptical arguments qualifies her as "dismissive," or as someone who "believe[s] that warming is not happening, is not a threat to either people or non-human nature, and strongly dispute[s] that it is a problem that warrants a national response."[13] By contrast, Julie and Craig seemed to know less about climate change (for example, they were unfamiliar with arguments about the historical variability of temperature). They would therefore fit best in the "disengaged" category, which applies to those who "haven't thought much about the issue at all [and] don't know much about it."[14] While they were all technically end-time believers, their different views about whether climate change exists and fits into the end-time scenario led to tension.

Given other comments Julie and Craig made during the discussion, it appeared that their lack of familiarity with the dimensions of the climate change debate was connected to their focus on the end times. In particular, their belief that society was irredeemable seemed to have encouraged them to withdraw. Julie had commented that she watched the Weather Channel frequently because "that's the only thing we can watch that's appropriate," suggesting she might not have watched secular news programs. Regarding their negative views of society, Julie said that "there are no morals," and Craig noted that "what is wrong to us would be considered right to [society], and what's right to [society] is considered wrong to us. It's two reverse worlds." That Craig viewed society as polluted with lax morals was underlined by his comment that during the end times "there's going to be a *whole* lot of people wiped away" (emphasis his). In terms of politics, Craig's and Julie's responses to the item on my questionnaire that asked about political views made it clear that they were not interested in becoming active. Julie wrote, "I don't care much about politics," while Craig wrote that "[I] dislike politics when they stray from the 'word of God.'" Given that they described themselves as extremely conservative (Craig also identified himself as a Republican, while Julie left the field blank), they would likely have been exposed to skeptical discourses about climate change had they been more engaged, but their withdrawal from society had seemingly insulated them from this knowledge.

Hybridizing Hot and Cool Millennialism

Although most of my informants fell into either the hot or the cool millennialist camp, a few seemed open but not committed to the idea that climate change might be a sign of the end times. These individuals hybridized hot and cool millennialist views in that they were typically skeptical that human activities were responsible for climate change, but also allowed for the possibility that if the climate was indeed changing then this might be a sign of the end times. That they wavered between being skeptical that climate change was occurring and wondering whether it was happening as a sign of the end times underlines the dynamic nature of hot and cool millennialist worldviews.

Pastor Roger of the Independent Baptist church exemplified this ambivalence. He was mildly skeptical about climate change, saying that "I don't think it's the threat to our global society that everybody makes it out to be." When I asked him whether he thought climate change was due to human activities, he equivocated again, saying, "I think it's more attributable to us, or just maybe things that we've done over a period of centuries," but he quickly reframed the discussion in religious terms, adding that a broad array of natural phenomena (he included hurricanes, earthquakes, and volcanoes) would qualify as anthropogenic, in the sense that they were caused by human actions—specifically, the sinful actions of Adam and Eve. He then shifted to the topic of end times, observing that "God has an overall plan" and suggesting that global warming might be a part of this plan: "I'm not so sure that some of the major catastrophic events, from ice melting to tornadoes and hurricanes and earthquakes that are happening all over the planet, the tsunamis,[15] that these are not some sort of apocalyptic signs that things are rapidly approaching the end." When I quizzed him further about his end-time beliefs, however, he gave the standard response that I heard in cool millennialist churches, which left open the possibility of either an immediate or a delayed return: "I believe in the imminent return of Christ, and that He could come at any moment. . . . But I don't know the day nor the hour." At the same time, he was clearly interested in the end times, for I noticed a full set of the *Left Behind* series on his bookshelves.

People like Pastor Roger were open to the idea that climate change might fit into an end-time scenario but seemed to be awaiting further evidence before committing to such an interpretation. It seems possible that as the world experiences more dramatic effects of climate change (making it harder to deny outright that the climate is changing), some in this camp might shift toward a hot millennialist interpretation.[16] In the meantime, their ambivalent position demonstrates the interpretive challenges that climate change poses for those who are aware of skeptical discourses about climate change but also on the lookout for signs of the end times.

In 2006, Bob Abernethy of the PBS show *Religion & Ethics News-weekly* interviewed Richard Land, a Southern Baptist pastor who both helped shape the Christian Right and has worked specifically on environmental issues. When asked to respond to E. O. Wilson's book *Creation: An Appeal to Save Life on Earth,* Land tentatively accepted Wilson's claim that Christians and environmentalists should work together. But he took Wilson to task for the latter's claim (discussed in chapter 1) that end-time believers were unmoved by the planet's deterioration. "I personally have never met an evangelical Christian who believes that," he protested. "I'm beginning to wonder if it's a mythic figure." Giving the cool millennialists' typical response, he added that "Jesus himself . . . says no man knows the hour, the day of His coming."[17]

If my own field research is any indication of broader patterns in end-time belief, Land was wrong. The figure is not a myth. Some evangelicals are sincerely convinced that the end is near and see climate change as a sign of what is to come. But in another sense Land was correct, for the end-time apathy hypothesis has attained a mythic status, one that did not seem fully supported from my time in the field. Indeed, the image of the end-time believer who exults at signs of planetary decline may only reflect the views of a small percentage of end-time believers. This seems even more plausible in light of research showing the difficulty of maintaining intense excitement about the imminence of the end times over time.[18]

This is not to say that end-time beliefs could not have such an effect under certain circumstances, particularly when a church or denomination is relatively disengaged from society. In fact, while I was in Georgia I spent time with a denomination that fit this very profile: the Jehovah's Witnesses. Drawn to attend their annual convention by its intriguing environmental theme—whether God would allow humans to ruin the earth through global warming and other environmental disasters— what I heard there vividly illustrated how end-time beliefs could encourage environmental indifference.

On a sweltering day in July, a speaker told a racially diverse audience of nearly seven thousand Witnesses, packed to the rafters in an arena-sized convention hall, that man-made pollution was "literally ruining the planet." This should not be cause for worry, however, because "Jehovah will not allow man to make this earth unfit for human habitation" (this line, repeated several times in the talk, inspired polite applause). Instead, after the earth had been purified and "kingdom rule" restored, "everything will be perfectly balanced again as it was in the beginning." At least in terms of the message being delivered (which may have differed from the message received), it was the end-time apathy hypothesis brought to life. Interestingly, there was no reference in any of the sermons that day to a responsibility to engage in public life. In fact, the speeches actively discouraged societal engagement, even telling a morality tale in which the heroine, a young African woman, forsook her dreams of higher education in order to spread the gospel door to door. If ever there was a blueprint for end-time-inspired passivity, it was here, where believing that the end times were near discouraged both concern about the planet's declining environmental health and any type of civic involvement in response.

But evangelicals have charted a very different path than the inward-looking Jehovah's Witnesses. Despite their belief that Jesus could return at any moment, evangelical Christians have, over the past half-century, gradually become ever more actively engaged with the society around them, with a particularly sharp focus on the role of Christianity in public life and politics. While it is not surprising that a few individuals in evangelical churches I visited were drawn to end-time beliefs, it makes sense that for the majority of my informants (the cool millennialists), end-time beliefs were not a daily preoccupation. Perhaps surprisingly, though, it was in fact this spirit of engagement—or, more precisely, *embattlement*—with the broader culture that guided interpretations of climate change most powerfully.

CHAPTER 4

The Embattled Mentality and Climate Skepticism

Although I did not know it at the time, I encountered a major clue about the role of the embattled mentality in fostering climate skepticism at the first church I visited during my fieldwork. The church inhabited a plain building adorned only with its name, Mayfield Church of Christ. Stepping inside offered a pleasant view of the nave, which was lit by a reddish light filtering through ceiling-length windows. Joshua, the pastor, greeted me and led me to his office, a kind of attic room that looked out over the nave. After I explained my project, he told me more about the Churches of Christ denomination: "We try to go through all the fog and haze of tradition and go back to the Bible. . . . Sometimes people ask, what does the Church of Christ teach about this or that thing? And I answer, nothing. The Church of Christ doesn't teach. The Bible teaches."[1] Although he did not want to label his church as belonging to a particular movement within Christianity, it was a quintessentially evangelical answer, capturing the tradition's high regard for scripture.

Over the next hour we talked about a range of topics, including Joshua's church, his faith, and his attitudes toward the environment and climate change. Like the majority of other pastors I interviewed, he did not believe that the climate was changing on a large scale, and he speculated that most other pastors in his denomination would agree. As the interview was winding down, I asked if there was anything else he felt I should know that related to our conversation. Nodding, he rummaged in his desk for a moment before extracting a shrink-wrapped DVD.

I glanced at the title: *The Silencing of God: The Dismantling of America's Christian Heritage.* I flipped it over and read the back: "Take a trip back to the now WITHERED ROOTS and vanishing values of America's past. You will not believe your eyes and ears. America is in the throes of a full-scale CULTURE WAR. *The moral and spiritual underpinnings of American civilization are collapsing*" (emphasis in original).[2]

Since Joshua claimed it would help me understand how the Christian worldview related to the environment, I made a note to watch it soon. But it had no obvious connection to the topic, and so I set it aside. It was not until months later that I would come across the DVD in a file folder and, with a better understanding of evangelicalism in place, recognize its basic argument as exemplifying the embattled mentality that made my informants suspicious of environmentalists and skeptical about climate change.

DISCOVERING THE ROLE OF THE EMBATTLED MENTALITY

Although the connection now seems obvious, understanding the link between evangelicals' sense of embattlement with secular culture and their attitudes toward climate change was a slow process. Indeed, my months of field research in the summer and fall of 2011 did not yield clear conclusions about the origins of these views, whether they were theological, political, or something else. Almost everyone I spoke with was skeptical about climate change, and many were hostile toward environmentalists.[3] But while it was clear that my informants' faith was shaping their environmental attitudes in some way, I could not pin down how, much less the origins of the connection. Not until I met with a group of Seventh-day Adventists early in 2012 was I able to see that there was in fact a pattern—one that, until then, I had been too close to see.

As mentioned earlier, Seventh-day Adventism is a tradition that has historically emphasized premillennialism.[4] According to the end-time apathy hypothesis, this should have made them strongly anti-environmentalist. Yet, minus the few hot millennialists I encountered there, my focus group actually revealed positive attitudes toward environmentalism and climate change—much more positive than those I had encountered in the evangelical churches I had visited. What accounted for the difference? As I pondered my experiences at the various churches, I realized that the Adventist focus group participants had stood out not only because of what they said, but also because of the manner in which they said it. At the evangelical churches, there was a kind of angry undertone to the conversations—especially, but not exclusively, when talking about the environment. With the Adventist focus group, however, the discussion had been calm, even gentle.

Not knowing exactly what to call it yet, I hypothesized that there was something in evangelical subculture that was shaping my evangelical informants' attitudes toward climate change. Whatever it was, it did not seem to be operating among the non-evangelical Seventh-day Adventists. Not long after, I happened upon the following passage in my notes from the 1998 book *American Evangelicalism: Embattled and Thriving* by Christian Smith and colleagues: "American evangelicalism . . . is strong not because it is shielded against, but because it is—or at least perceives itself to be—embattled with forces that seem to oppose or threaten it. Indeed, evangelicalism . . . *thrives* on distinction, engagement, tension, conflict, and threat."[5] I immediately recognized this portrait. Like the evangelicals from around the country that Smith and his colleagues had interviewed, my informants were angry because of the secular forces they saw arrayed against them, and because of the things that they felt had been taken away from them. From my focus groups, it was clear that my informants viewed climate change through the same lens. For them, climate change was not only—as secular climate skeptics might say—a hoax based on weak science; it was also a tool wielded by secular elites to undermine the Christian worldview.

ENCOUNTERING THE EMBATTLED MENTALITY

To those familiar with the history of discrimination against racial and ethnic minorities in the United States, the idea that Christians—a group that has dominated the country's history and still claims adherents among the majority of the population—face discrimination and persecution seems counterintuitive. Yet this is exactly the view that many evangelicals today hold. According to a 2016 study, for example, nearly eight in ten white evangelical Protestants felt that discrimination against Christians rivaled that experienced by other groups.[6] Such concerns have been amplified via conservative and religiously conservative news websites like Townhall.com and WND.com, which highlight the narrative of discrimination via stories such as "7 Examples of Discrimination against Christians in America" and "Attacks on Christians in US Double in Three Years."[7] Even though it is a central part of many evangelicals' worldview, however, few have grasped the impact of this idea on evangelicals' attitudes toward the environment and climate change.

At the evangelical churches I visited, the power of the embattled mentality to shape how its adherents viewed the world was evident from the outset. As a way of breaking the ice in the focus groups, I opened each session by asking participants two very general questions: what they felt Christianity contributed to society, and what challenges Christians faced. The goal was simply to establish a common ground for conversation, but the questions ended up giving me a critical window into my informants' world and, in turn, their perspective on climate change.[8]

To summarize their responses, my informants tended to describe the church as existing in its own sphere, distinct from the larger society (a society that evangelicals often refer to simply as "the world").[9] They described their churches as safe havens that offered people things that they desperately needed but that were lacking in society at large, such as moral and spiritual guidance, hope for the future, and material assistance. At the nondenominational Christian Church I attended, for example, Kelly said—to agreement from the group—"to me the greatest contribution that Christianity makes is hope. It gives people hope for new beginnings, for forgiveness, for . . . something to live for." Kim, a woman in her twenties, agreed, adding that "I think [what Christianity contributes is] love and support. When you see people who aren't Christian coming to church, it's when something bad has happened in their lives, and they don't have people in society to love them so they'll come to church and then they find it there." Highlighting the contrast with secular offerings, Graham, who attended a Southern Baptist church, played off of Barack Obama's 2008 campaign slogan, "Hope and Change":

> To piggyback on a saying that was used last year, Christians provide hope and change: a change in life from dire situations, and a hope for a better future, and that in doing good for others that they will in turn be blessed. So Christians provide *true* hope and *true* change.[10]

Others mentioned missionary work, moral guidelines, and love as benefits that Christianity provided. Many of my informants also emphasized that the church was like a family to them, a metaphor that seems apt when one considers the degree of emotional, spiritual, and material support the church offered them.

Despite these important contributions, however, my informants felt that mainstream society self-defeatingly rejected what their faith had to offer. In particular, I heard many references to how Christians were being marginalized, ridiculed, and persecuted. In the Presbyterian focus group, for example, Lee discussed the challenges that Christians faced when interacting with the larger society, stating that

> the Bible talks about just as He [Jesus] suffered and was spit on and made fun of, it guarantees that you as Christians will go through the same persecution. I think that's the reality of living in the world—*in* the world but not *of* the world—that you're not going to be the most popular group.

The feeling of having to bear scrutiny because of one's faith was widespread.

Another set of complaints focused on schools, which many of my informants saw as sites from which faith had been expelled via the removal of prayer and the insistence on teaching evolution. At the Southern Baptist church, for example, when I asked about challenges that Christians faced in society, Shirley, a teacher in her forties, responded that

> my first thought is persecution. Christians are persecuted. I am a teacher . . . and of course being in a public school, we're not allowed to mention God. We do have a pledge of allegiance and moments of silence and the students just stand there and don't know what the purpose is, and they don't really know what it's for. It's sad that we can't somehow bring that into their education.

A number of informants used this same example to illustrate how Christianity was under attack.

In a separate interview, Shirley's pastor elaborated on this theme, arguing that "the whipping boy today is the real Bible-believing Christian." Pastor Richard's example was the football quarterback Tim Tebow, a devout evangelical Christian who became famous as a college football player and subsequently had a brief professional football career (he has since moved on to baseball). He was hugely popular among evangelicals, who valued his regular and sincere expressions of faith. As evidence that Christians were being persecuted, the pastor cited a pro-life television commercial that Tebow had recently made with the evangelical parachurch organization Focus on the Family.[11] According to Richard, "[Tebow] did just a simple commercial thanking his mom for not aborting him. Well, that was like spitting in the eye of the crowd." As he explained, Tebow's example illustrated that Christians who stood up for what they believed in faced societal condemnation, because "when you stand for what you believe, you will be out of step with the world. . . . Christians can easily become hated because of our adherence to Christ's teaching and who we believe him to be historically."

Like Pastor Richard, many of the Christians I spoke with felt that they were being pushed out of society, marginalized, excluded, and persecuted because of their adherence to their faith. Critically, they felt this to be a departure from the historical norm, which had established America as a Christian nation. For example, Anne, a health-care worker in her forties who attended the Christian Church, asserted that

> I think we should be real concerned about them pushing and stripping God and Christianity, religion out of politics. We came over to this country to begin with for freedom of religion, but we're supposed to be separate now [i.e., the wall of separation between church and state], so they're going back and changing all the things that we started this country about.

In her view, America's Christian heritage was in the process of being dismantled by those who sought to remove Christianity from public life. As Christian Smith's work shows, my informants were well within the evangelical mainstream in expressing these views.[12]

Of course, it is important to remember that evangelicals are a diverse group, and not all of them share this perspective. As Smith and his colleagues have pointed out, other stories could be told about evangelicals; this is just one of many facets of their experience and identity.[13] But at the same time, it was no coincidence that this perspective came up at every evangelical church I visited, for it is one that has deep historical roots.

HISTORICAL ORIGINS OF THE EMBATTLED MENTALITY

Evangelicals' sense of embattlement with secular culture has roots in the late 1800s. Before the modern evangelical movement had emerged, many Protestants felt threatened by emerging intellectual trends. As mentioned in the introduction, chief among these were new techniques of biblical criticism and Darwin's *On the Origin of Species* (1859). While some Protestants were eager to reinterpret their faith in light of the new scholarship and scientific research, those who resisted increasingly adopted a siege mentality. "Never since the crucifixion," lamented the president of Amherst College in 1870, has Christianity "been assailed with such variety and persistence of argument."[14]

This sense of opposition to secular culture grew as the fundamentalist movement emerged in the early twentieth century, and it became more entrenched as fundamentalists built up a vast infrastructure—churches, denominations, Bible institutes, colleges, seminaries, Bible camps, mission societies, and publishing houses—to aid in defending the faith. Not only did fundamentalists oppose intellectual trends, they also resisted other societal changes during this period, defending traditional gender and family roles as progressive women agitated for birth control, education, and voting rights.[15] By the time of the Scopes Trial in 1925—a highly publicized court case that tested a law against teaching evolution in public schools—fundamentalists already saw themselves as "besieged outsiders who were increasingly at odds with a culture rapidly ceding ground to the forces of the Antichrist."[16]

While fundamentalists were concerned about numerous groups that they believed threatened America's Christian heritage (chiefly Catholics and Jews), they were also beginning to be particularly critical of a new threat—secularism. Higher education was a particular target, in part because it represented territory only recently lost: until the late nineteenth century, most such institutions had been headed by Christians who believed firmly in using the faith's teachings to guide students' moral and spiritual development.[17] Now, Christianity was "simply another lens through which to view the world."[18]

During the interwar period, fundamentalists complained relentlessly about the "Godless faculties" populating America's colleges and warned that the "God-denying, Christ-reputing, Bible-scorning theories" promoted in modern universities would lead to a Soviet "triumph in America" and "the overthrow of the Christian state."[19] Public schools were no better, being subject to most of the same secularizing trends. As William Jennings Bryan had argued in 1921, "atheists and agnostics are not only *claiming* but *enjoying* higher rights and greater privileges in this land than Christians; that is, they are able to propagate their views at *public expense* while Christianity must be taught at the *expense of Christians.*"[20]

Fundamentalists were concerned not only about secularism but, like other Christians of the era, about increasing state power. As the historian Kevin Kruse has shown, during the 1930s, business interests were seeking a way to restore the public's faith in free enterprise, which had been deeply damaged by the stock market crash and ensuing Great Depression. They eventually adopted the strategy of using prominent Christian pastors to help disseminate their message, via the argument that the New Deal's "collectivism" was a "glorification of the state" that would encourage Americans to worship the government rather than God.[21] Although the message was spread to a wide audience of Christians, fundamentalists had strong ties to the business community and were especially receptive to the idea. As one fundamentalist author of the period put it, it was their Christian duty to "'abolish the socialist bureaucratic encroachments' on constitutional government, [and to] fight to conserve 'our Christian heritage.'"[22] Hence, in the interwar era, fundamentalists became concerned not only about secularism (along with Roman Catholicism and theological liberalism), but about "statism," or dependence on the federal government encouraged by an excess of social welfare programs.[23] Fears about secularism and statism would become particularly important in the era of climate change, when climate change advocates would be portrayed as both allied with secularists and working in the service of big government.

Concerns about secularism and statism were both evident at the inaugural meeting of the National Association of Evangelicals (NAE) in 1942. In his inaugural address to the NAE, president Harold John Ockenga had warned of "ominous clouds of battle which spell annihilation unless we are willing to run in a pack."[24] In his view, the clouds came in the form of secularism and the growth of government under the New Deal—but also in the form of Roman Catholicism and theological liberalism. Several changes would have to occur in the broader society to convince evangelicals that secularism was the biggest of the battle clouds and that, to address it, they needed Roman Catholics as allies rather than foes. The first had to do with the nation's public schools.

While conservative Protestants had been decrying the secularization of education since the early twentieth century, a series of Supreme Court decisions in the 1950s and '60s raised the stakes considerably. During the first half of the century, the extent to which religion was permitted in public schools was a matter of state law. Some states (primarily in the Midwest and West) forbade religious observances in public schools, while others required them. The reason for this was that the First Amendment—"Congress shall make no law respecting an establishment of religion or prohibiting the free exercise thereof"—was then interpreted as applying only to laws that *Congress* would enact. States were thus left free to regulate religious practice as they pleased. After the ratification of the Fourteenth Amendment in 1868, however, the Supreme Court decided that the First Amendment applied to states as well. This ultimately led the Supreme Court to hand down a series of decisions in the 1950s and '60s that prohibited religious practices like Bible reading and prayer in public schools because they violated the First Amendment.[25] Rather abruptly, a practice that had been regarded by Christians around the country as essential to preserving the nation's good moral character became unconstitutional.

The removal of prayer from schools also proved to be a catalyst dividing evangelicals from their mainline Protestant brethren, while also drawing them closer to conservative Catholics. While the two groups certainly already differed on many issues, during the 1950s ministers from both mainline and evangelical traditions had supported the notion that America was a Christian nation, enthusiastically supporting Eisenhower-era legislation to add "under God" to the Pledge of Allegiance and "In God We Trust" to the nation's currency. When the Supreme Court concluded in *Abington School District v. Schempp* (1963) that compulsory Bible reading violated the First Amendment, however, mainline and evangelical denominations diverged sharply in their responses. Mainline churches readily made peace with the new ruling, arguing that "the real threats to faith were the 'common-denominator religious exercises' struck down by the court."[26] But evangelicals were incensed. In forbidding the reading of scripture itself—the centerpiece of the faith from the evangelical perspective—*Schempp* had crossed a line. According to the NAE's president at the time, the decision was "a sad departure from this nation's heritage under God."[27]

While unpopular, the court's decision did not immediately lead evangelicals to mobilize because they worried that opposing the decision would give Catholics a political advantage, and because school districts around the country widely flouted the decision.[28] By the 1970s, however, things were different. In surveying the changing culture around them, with increasing acceptance of abortion and gay rights, support for feminism, and the introduction of sex education into public schools (soon after the Bible was taken out), many evangelicals became convinced that the era's moral decline stemmed from a singular source: secular humanism.[29]

Events in the political sphere would soon serve to magnify this perception. By 1976, the New Right activist Paul Weyrich—a Catholic who saw the potential of uniting social and economic conservatives under the umbrella of opposition to big government—began working to harness evangelicals' growing discontent for the purposes of political mobilization. Working toward that end, the Heritage Foundation, his brainchild, would publish *Secular Humanism and the Schools: The Issue Whose Time Has Come* (1976).[30]

Boosting the efforts of New Right activists such as Weyrich was the evangelical philosopher and theologian Francis Schaeffer, who became a key player in translating the idea of secular humanism into a force for political mobilization. In his 1976 book *How Should We Then Live?*,

Schaeffer argued that Western societies had replaced God's moral stand-
ards with secular humanism, leaving them in a kind of moral free fall.
His book, which became a best seller among evangelicals, gave evangelicals a framework for understanding the societal changes around
them. It also helped galvanize them against abortion, which was one of
Schaeffer's primary examples of the disastrous results of secularism.
While Schaeffer had initially been nonpartisan, his opposition to abortion eventually led him to partner with leaders like Jerry Falwell and Pat
Robertson, thus helping to build the new Christian Right coalition.[31]

Ironically, given the way his campaign against secular humanism
played out, Schaeffer was also one of the earliest evangelicals to advocate environmental concern. In *Pollution and the Death of Man* (1970),
he decried the world's growing ecological problems and argued that if
properly understood, Christianity was uniquely suited to promoting
concern for the earth. In fact, relying on the same distinction that he
later articulated in *How Should We Then Live?*, he argued that Christians could be better caretakers of the earth than secularists:

> Secular man may say he cares for the tree because if he cuts it down his cities
> will not be able to breathe. But that is egoism, and egoism produces ugli-
> ness. . . . [In contrast] the Christian stands in front of the tree, and has an
> emotional reaction toward it, because the tree has real value itself, being a
> creature made by God.[32]

Conscious of an evangelistic opportunity, he urged action: "We must
exhibit that, on the basis of the work of Christ, the Church can achieve
partially, but substantially, what the secular world wants and cannot
get."[33] While the book would have appealed to young evangelicals concerned about the environmental crisis, it had far less impact on the tradition as a whole than did Schaeffer's discussion of secular humanism
and anti-abortion activism.

With Schaeffer's identification of secular humanism as conservative
Christians' primary foil, it was easy to see (or claim) in retrospect that
the change in federal policy regarding Bible reading in public schools
had catalyzed the movement. For example, in the telling of Harry Jackson Jr. and Tony Perkins, two evangelical leaders now active in the
Christian Right, the moment was foundational to the emergence of the
Religious Right:

> The religious Right is made up of people who have been awakened by real
> concern over governmental policy and cultural initiatives that were being
> influenced and shaped by a postmodern worldview that was radically secular

and hostile toward the Christian faith. The specific galvanizing issue that launched and rapidly advanced the cause of the Moral Majority was school prayer, which had been banned by the federal courts. These courts had become a virtual candy store for people promoting the radical secular agenda. Stripping public schools of prayer ... was the issue that finally caused millions of Christian voters to march back into the public policy arena.[34]

In fact, the secularization of public schools was just one of many factors that led to the movement's emergence. Leaders like Falwell and Robertson have publicly cited anti-abortion activism prompted by the Supreme Court's *Roe v. Wade* decision in 1973 as pushing them into politics. There is little doubt this became a major driving factor for lay evangelicals (with the encouragement of right-leaning Christian political organizers). For leaders in the movement, however, it was the federal government's threat to deny tax-exempt status to segregated Christian schools that activated them.[35] Feminism and gay rights also provided useful foils for newly engaged evangelical activists.[36]

While multiple factors contributed to the political mobilization of evangelicals, there is no doubt that the vastly unpopular decisions related to Bible reading in public schools were helpful—as was evident in the frequency with which prayer in schools came up in my own interviews (humanism came up only in interviews with pastors, who could generally put a finer point on the sources of their discontent). If evangelicals began to feel embattled by secularists in the 1960s, by the late '70s they were beginning to feel emboldened—to reclaim their heritage as moral guardians of the nation, and to return the Christian faith to its rightfully central place in American society. Anti-secularist discourse would play a key role in this project.

Since the 1970s, a growing chorus of voices in both the political and Christian Right have contributed to this discourse.[37] "It is no overstatement to declare that most of today's evils can be traced to secular humanism," warned the prominent evangelical author and leader Tim LaHaye in his 2000 book *Mind Siege*. Indeed, humanists had "already taken over our government, the United Nations, education, television, and most of the other centers of life."[38] Other such voices include the Reverend David Limbaugh (brother of the conservative political commentator and radio personality Rush Limbaugh), James Dobson (founder of Focus on the Family), the evangelical politician and activist Gary L. Bauer, and Bill O'Reilly, also a conservative political commentator.[39] Indeed, the evangelical mass media (which include radio, television, and digital media) have also become "a vast 'echo chamber' where the Christian nation thesis is nurtured among evangelicals without competition from other ideas."[40] With so many high-profile figures disseminating this message, it is hardly surprising that it eventually came to resonate at the lay level as well. In fact, leaders would have stopped deploying this discourse if it did not resonate. Its very persistence speaks to its ability to motivate political activism and engagement.[41]

While not explicitly political in its content, the discourse of embattlement does have a crucial political function: it sustains evangelicals' sense of identification with the Republican Party. That is, in pitting them against liberals, both political and theological, it naturalizes evangelicals' identification as Republicans.[42] The blending of that religious and political outlook was clearly evident in the way my informants talked about the environment and climate change.

HOW THE EMBATTLED MENTALITY UNDERGIRDS CLIMATE SKEPTICISM: RELIGIOUS SOURCES OF SKEPTICISM

My informants expressed a number of common rationales for skepticism that were rooted in their faith. Examining these rationales shows the extent to which my informants viewed climate change through the lens of embattlement, as a threat to their faith that had been concocted by secularists and their allies as part of a broader campaign to diminish Christianity's influence in American society. From this perspective, rejecting climate change went hand in hand with defending their faith.

"God Is in Control"

The most common way in which it was apparent that the embattled mentality shaped attitudes toward climate change was in my informants' insistence that human activities did not, and indeed could not, affect the climate. From their perspective, the idea that humans could alter the planet was simply a counter-narrative designed to undermine Christian teachings about divine omnipotence. For example, Peter from the Independent Baptist church stated that

> when I look at climate change . . . I have the mindset that God's in control. If He doesn't want it to get hot it's not going to be. And I know a lot of people think that's a naive attitude. But I think that's where we go back to, if He doesn't want it to happen it's not going to happen.

Even knowing that others might perceive his attitude as "naive," Peter held firm in his faith—an act that becomes particularly meaningful, if not courageous, in the context of the belief that Christians were being marginalized and excluded from society.

For Stephen from the Christian Church, climate change fit into a larger pattern of hubris that was typical of environmental narratives:

> At the root of [environmentalism] is the belief that man is in control of this world. . . . For instance, the experts will debate with global warming . . . [whether] the climate will get so hot that we'd melt the glacier ice caps. . . . Yet nowhere in the discussion is the possibility that God might be able to maintain a consistent temperature. I think at the core of it is the denial that God is in control, and that the earth's destiny is within the hand of man, rather than God.

According to Stephen, a fundamental problem with global warming was not scientific but theological: the narrative of climate change ignored God's power to control the weather as well as, more broadly, earth's destiny. Regardless of expert opinion, climate change was objectionable because it was premised in a worldview that sought to erase God and deny his sovereignty. More to the point, climate change was not an actual threat but one concocted for the express purpose of undermining Christian teachings.

Participants at other churches echoed this perspective by suggesting that it was not only inaccurate to think humans were in control of the climate, but arrogant. Lee from the Presbyterian church, for example, rejected anthropogenic climate change because in his view humans were "playing the role of God by saying it's all our fault." Similarly, Tyler from the Church of Christ argued that "if you take God out of the equation, global warming is your least concern." In other words, with climate change, as in schools and other areas of public life, Christians were under assault by those who wanted to deny God's power. Viewed through the lens of embattlement, climate change appeared to be part of a pattern of marginalization whose real purpose was to reduce Christianity's role in public life.

Environmentalists as Secular Elites

Evangelicals who believe that Christianity in America is under attack have identified a number of groups as responsible for the changes they lament. Liberals, Hollywood, elites, feminists, the media, and academics have all been accused of helping to advance a secularist agenda.[43]

According to the discourse of embattlement, these "secular elites" are using all the tools available to them to shift American culture away from its biblical foundations.[44] As I discovered, my informants viewed environmentalists and climate change activists as allied with these secular elites.

The most obvious way in which my informants associated environmentalists with secular elites was in their frequent assertion that environmentalists were trying to take God out of the equation, just like other secular elites. In a few cases, however, my informants explicitly associated environmentalists with other secular elite groups. Pastor Richard from the Southern Baptist church suggested that the reason evangelicals did not typically embrace environmental issues had to do with their provenance:

> You'll have an idea come out of Hollywood or out of the world, and it'll be so far to the left, that we [Christians] jerk to the right. . . . So there are people who don't have our leanings or our beliefs in who God is and morality, and they seem to be the ones who champion this whole idea of the stewardship of the earth.

Here Richard not only pointed out that there were differences in opinion about the environment, but that specific social groups—"Hollywood," "the world," or people who had different ideas about faith and morality—promoted environmentally friendly views, and this was part of what discouraged evangelicals from being concerned about the environment. That he associated environmentalists with Hollywood is particularly noteworthy, given the frequency with which politicized evangelicals identified Hollywood as a cause of America's moral decay.[45]

Similarly, Pastor Roger from the Independent Baptist church drew a clear line between those whom his fellow Georgians admired and the (conflated) Hollywood/global-warming crowd:

> [This part of] Georgia and small towns are pretty unique in being very conservative, if I can use the word *anti-Hollywood*, [while] a lot of what we might term the "global warming crowd" is typically—they're not the people that a lot of folks in [this part of] Georgia and small towns typically look up to.

Here Roger linked a secular elite group, the Hollywood crowd, with climate change activists, implying that he was suspicious of both for similar reasons. It was clear from the kinds of statements my informants made about environmentalism that many of them viewed it as an ideology designed to undermine Christianity.[46]

Powerlessness and the Importance of Not Worrying

My informants' emphasis on God's omnipotence also led them to emphasize that humans were powerless to affect the climate—and, moreover, that as Christians they ought not to worry about such things. As with the assertion that God is in control of the climate, believing in climate change was thus framed as demonstrating a lack of faith. Conversely, my informants presented not worrying as a way of affirming their Christian worldview—yet another way in which their focus on defending their faith weakened the potential for concern or action. According to Michelle at the Independent Baptist church, for example:

> I don't think that anything that I do is going to change the outcome, because He already knows what's going to happen. So I just need to get up every day to trust Him to give me the sense to take care of it. And that's exactly what I do.

Determining what exactly is God's will and what is not is a tricky business, however, and my informants rarely achieved clarity on this point. Generally their solution was that they should do what they could on a local level and assume that God is in charge of the big picture. According to Kara, a nurse in her forties who attended a Presbyterian church, for example:

> Yes, we do need to quit using aerosols and things that we can do to try to help it. But if it is going to happen it's going to happen. We can try to prevent it, we can try to slow it down, but ultimately things are going to play out the way God wants them to.

For Kara, if God was in control, little could (or should) be done to prevent the outcome he desired. Mark, a church administrator in his fifties who attended the Southern Baptist church, reasoned through the conundrum in a similar fashion. He recalled how lead in paint used to cause health problems, yet this problem was eventually "fixed through man's actions, I believe because God wanted it fixed." Referring broadly to current events, he concluded that "if it's part of God's plan it's part of God's plan. But if it's to be fixed, He'll use His people through their giftedness to fix it. And if God's there, wherever God is, His will be done." His logic failed to clarify under what circumstances the onus is placed on humans to take initiative to solve problems. I heard such arguments frequently, and probing further did not result in any resolution. Regardless of its illogic, the notion that God is in control was highly effective for justifying complacency regarding the planet's deteriorating condition.

My informants' faith also shaped their attitudes toward climate change via their reluctance to express worry, not just about climate change but about life in general. In particular, many of my informants argued that worrying indicated a lack of faith in God. At the Christian Church, for example, Kelly argued that to worry about climate change was pointless:

> I'm not worried about whether we're going to be hit by a meteor or if we're going to have global warming. To me, I think we should take care of the earth, I think He gave it to us to take care of, but to sit and worry about all the things that will happen to me is taking God out of the equation.

For Kelly and many others, to worry about climate change was to deny God's centrality as an active force in their lives. Doing so, moreover, was dangerous because it would play into the hands of secularists.

My informants' insistence that they should not worry made sense in light of Jesus's famous injunction in the Sermon on the Mount to "take therefore no thought for the morrow, for the morrow shall take thought for the things of itself."[47] Indeed, this teaching had much the same effect on attitudes toward climate change that sin and salvation had on my informants' attitudes toward environmental problems: it reduced urgency and reframed concern in a context that was more spiritually meaningful. But while my informants had otherwise affirmed the importance of maintaining a clean environment, climate change was in this case reduced to nothing—a source of anxiety that was both unnecessary and spiritually unwise.

Climate Change as a Competing Eschatology

The final way in which the embattled mentality shaped my informants' attitudes toward climate change is of particular interest given the origins of this study. Many of my informants denied that climate change was occurring on the grounds that God is in control of the end times. As Tyler put it, "Regardless what we do or don't do, the world is going to come to an end. It's going to be, in accordance with the scriptures, it's going to be burning up, it's going to be destroyed—by God, not us." In Tyler's view, it was arrogant to suggest that humans could destroy the world; this was another attempt to deny God's power. In almost exactly the same words, the pastor of a nondenominational evangelical church in Dixon argued that according to his understanding of scripture, climate change could not be catastrophic:

I'm not a scientist. But I do know this: I know that going back to the scripture, I know that God promises He's never going to flood the earth again and destroy it. The next time He comes it'll be different. . . . I know how the story ends. It doesn't end with us drowning in glaciers, or anything like that.

As these quotations illustrate, to the extent that end-time beliefs were affecting the majority of my informants' attitudes toward climate change, it was not because they did not care about protecting the earth, owing to their conviction that it would soon be destroyed (as was the case with hot millennialists); rather, it was because their sense of themselves as defending America's Christian heritage, which manifested itself here through their commitment to orthodox Protestant teachings about the end times, led them to reject a phenomenon that they believed was intended to undermine that worldview.

This connection is more politically potent than the one supposed by the end-time apathy hypothesis. My hot millennialist informants' attitudes toward climate change were hitched to an engine of this-worldly apathy: If this world will soon pass away, why get involved in politics? But my cool millennialist informants' attitudes were hitched to an engine of this-worldly engagement: Christianity in America is under attack and must be reclaimed for the faithful.

This distinction also speaks to a broader methodological point. Indeed, I believe that a methodological problem underlies the end-time apathy hypothesis, one that Barney Glaser and Anselm Strauss, the architects of grounded theory, called "exampling." "A researcher can easily find examples for dreamed-up, speculative, or logically deduced theory after the idea has occurred," they wrote in their groundbreaking book on the methodology.[48] But, Glaser and Strauss continue, since the hypothesis was not derived from the examples—rather, the examples were selected on the basis of the hypothesis—there is no chance for the examples to change the hypothesis. That is, there is no opportunity for the data to alter or better inform the theory. As a result of this "opportunistic" use of theory, Glaser and Strauss conclude, "one receives an image of proof when there is none, and the theory obtains a richness of detail that it did not earn."[49]

In the case of the environmental impact of end-time beliefs, observers have for decades brought the end-time apathy hypothesis to life by citing James Watt and a few others. But the absence of open-ended field research has afforded little opportunity for the data to alter or better inform the theory. Conducting inductive field research yielded a different interpretation, one that still acknowledges the importance of end-time beliefs but that is more richly grounded in the data of lived experience. Like those who favor the end-time apathy hypothesis, that is, I recognize the significance of prophecy belief for evangelicals. But spending time in evangelical churches led me to place the emphasis elsewhere when it comes to understanding environmental attitudes. For most of my evangelical informants, end-time beliefs themselves did not simply and directly produce environmental apathy; rather, it was the broader complex of beliefs and attitudes toward the growth of secularism in American culture that produced righteous indifference, or even animosity, toward those who warn of human-induced environmental decline.

It is important to add that because the intensity of interest in the end times can change dramatically over time, the embattled mentality may not always be the best explanation in the future, and indeed it may not be the best explanation of previous historical moments. But it seems to explain well what I encountered in Georgia and—as I will argue later—within the United States more broadly.

EVANGELICALS, POLITICS, AND THE CLIMATE DENIAL MACHINE: SECULAR SOURCES OF SKEPTICISM

The social scientific literature on climate change attitudes has established partisanship as one of the most powerful drivers of skepticism.[50] Specifically, Republicans and political conservatives are much more likely to be skeptical of climate change than Democrats and political liberals, and the gap has been growing over time.[51] My informants were overwhelmingly politically conservative, providing a useful opportunity to examine how political conservatism interacts with religious conservatism to shape skepticism. What I found was not only that my informants were conversant with many of the talking points advanced by those promoting climate change skepticism but also that many of these arguments resonated with them because of their religious outlook. Hence, secular and religious arguments for climate skepticism buttressed each other.

Connections between Conservative Politics and Climate Skepticism

One name given to describe the campaign to undermine the public's acceptance of climate science is the "climate denial machine." The term was first employed in a 2007 *Newsweek* article describing how, beginning in the late 1980s, a "well-coordinated, well-funded campaign by contrarian scientists, free-market think tanks and industry [created] a paralyzing fog of doubt around climate change."[52] According to the sociologists Riley Dunlap and Aaron McCright, the machine's strategy is "to undermine the case for climate policy making by removing (in the eyes of the public and policy makers) the scientific basis for such policies—i.e. by challenging the reality and seriousness of climate change."[53]

Borrowing from techniques developed by the tobacco industry to forestall government-imposed restrictions on their products, the climate denial machine has manufactured doubt through a variety of actors.[54] The fossil fuel industry, other energy- and resource-based corporations, manufacturing companies, and national business associations have all, for example, funded contrarian scientists, conservative think tanks that promote climate denial, and front groups that advance and publicize such positions. Wealthy conservative philanthropists have added to the flow of money, particularly by donating to conservative think tanks committed to promoting free enterprise and limited government.[55] Unlike the more self-interested corporate funders, think tanks tend to oppose climate science and policymaking for ideological reasons.[56] They are particularly effective at doing so because they have been able to present themselves as objective sources of information, a framing whose success has enhanced their credibility with the public.[57]

The climate change disinformation campaign has been especially successful among Republican voters for two reasons. First, Republican lawmakers in Congress have overwhelmingly opposed action (and in many cases have dismissed the scientific evidence itself). A 2007 poll of Congress found, for example, that 95 percent of Democratic respondents but only 13 percent of Republican respondents agreed that "it's been proven beyond a reasonable doubt that the Earth is warming because of man-made problems."[58] Research on the formation of public opinion has shown that for complex issues like climate change, citizens tend to take shortcuts rather than engage in extensive research themselves.[59] Since one of the simplest shortcuts is to follow the cues of political elites, Republican voters have ended up being more likely to be skeptical of climate change than Democratic voters because they have followed the cues of their political elites.[60]

Cues from Republican politicians play a major role in encouraging skepticism, but the media also play an important role in publicizing these cues.[61] This is especially so because of the "balkanization" of the

American news media, in which Americans tend to consume news from outlets that reinforce their preexisting political beliefs.[62] When it comes to climate change, conservative and liberal news media have given vastly different accounts of climate change. A content analysis of climate change coverage on Fox News, CNN, and MSNBC found, for example, that Fox took a more dismissive tone and interviewed a greater ratio of climate change doubters to believers. Accordingly, when viewers are surveyed, watching Fox News correlates negatively with acceptance of global warming, even after controlling for other factors.[63] Thus, within the climate denial machine, elite cues from conservative politicians and media coverage with a bias toward skepticism are two powerful means by which political conservatives have been encouraged to view skepticism as the more reasonable position on climate change.

The Impact of the Climate Denial Machine

The arguments that my informants used to dismiss the reality or seriousness of climate change suggested that the climate denial machine had shaped their views in important ways. To begin with, my informants were almost universally skeptical that climate change was caused by human activities. This view is not reflective of the scientific consensus, but rather of skeptical messaging on the subject.

The most common explanation my informants gave for their doubt, which I heard in eight of the nine focus groups, was that any changes to the weather were due to natural variation, rather than to human activities. For example, Rhonda from the Independent Baptist church attributed warming trends to natural cycles:

> I'm old enough to understand that I was in California in the seventies and it was a big movement that we were having another ice age because from the 1950s on there had been cooling trends. Isn't it cyclical? Aren't there natural cycles? Aren't there solar flares? Aren't there things that happen that run in cycles? So why wouldn't this be another cycle?

Many of those who believed that changes to the weather were due to natural variation referenced commonsense understandings of the weather as naturally capricious to counter the argument that recent extremes were due to climate change.

My informants often presented generically skeptical statements like these alongside claims that are more specifically associated with the climate denial machine. First, many informants argued that there was no scientific consensus about whether climate change was occurring, a false claim that the climate denial machine has circulated widely.[64] Another way in which the influence of the denial machine was evident was in the near-universal disdain my informants expressed for Al Gore. His name came up more than any other figure in discussions, and never in a positive light. Well known not only as a liberal politician but as a prominent climate activist, Gore has been a favorite target of conservative climate skeptics, especially Rush Limbaugh. In fact, a content analysis of conservative newspaper columnists' discussion of climate change between 2007 and 2010 found that they referenced Al Gore almost twice as often as they discussed any other topic over the period of analysis.[65]

Such efforts seemed to have hit home with my informants. When I asked the group at the Independent Baptist church what they thought about the scientific research on climate change, for example, my question elicited this mocking exchange:

Rhonda: Where has Al Gore been lately?

Peter: [jokingly] Inventing the Internet! I got it, I got it. Inventing the Internet.

Rhonda: You know, he was such a proponent out there in the news. . . . I think with him it was scare tactics. I'm not scared. And I think they're scaring everybody, I really do.

The frequency with which my informants referred to Al Gore is another sign that their opinions had been shaped by the climate denial machine.

The second most common way that my informants expressed skepticism was to question the motives of those who were attempting to move climate change onto the national agenda. This reflects another strategy common to elite-driven countermovements like that associated with climate denial. Known as "adversarial framing," this strategy involves using a variety of unsavory tactics—ad hominem attacks, finger-pointing, name-calling, and character assassination—to advance the countermovement's position by questioning opponents' trustworthiness.[66] Questioning opponents' motives offers multiple advantages because it forces opponents to spend time and resources countering the claims, shifts the conversation away from substantive issues, and engages a greater audience by making the conversation itself more dramatic and entertaining.[67] From the frequency with which my informants questioned the motives of climate scientists and activists, it was clear that such tactics had been particularly effective with them. The following exchange from the focus group at the Southern Baptist church was typical:

> *Graham*: Global warming has taken on a life of its own and in a lot of areas it's taken on a life for monetary gain for people. You know a lot of people [say], "Well, let's do this because of global warming," where in essence it's just greed. It's a way for them to make money.
>
> *Robin*: Like businesses or scientists?
>
> *Graham*: I'd say mostly politicians.
>
> *Shirley*: I was going to say, it's a way to get elected. If that's a big issue in your life and a politician is saying, "Here's my plan to prevent some of these," then a politician's going to get elected, strictly based on a hot topic.

In addition to politicians, climate scientists were frequent targets as well. Most commonly, people suggested that these scientists were biased toward conducting studies that supported the theory of global warming because it benefited them financially. Stephen from the Christian Church articulated this suspicion succinctly:

> I'm not convinced that it [climate change] is even happening. And I go back to two questions guiding that. Is there good money being made in the perpetuation of the studies? [murmured agreement from group] And would they continue to make money if science discovered that it's not that way?

In support of this theory, a number of informants referenced the "Climategate" scandal, in which emails stolen from climate scientists at the University of East Anglia in the United Kingdom (a group that was working with the United Nations' Intergovernmental Panel on Climate Change) showed that the scientists had discussed how the science supporting climate change should be portrayed. Climate skeptics subsequently used this information to argue that climate scientists were, in a corrupt manner, manipulating the data to support their case, rather than conducting solid scientific research. Aided by a well-orchestrated public relations campaign, news of the "conspiracy" spread quickly. Even though none of the charges of corruption or manipulation held up, and lengthy investigations by the associated universities concluded that no research misconduct had occurred, many of my informants cited the controversy in describing their reasons for not trusting climate scientists.[68] Their evident familiarity with this adversarial framing strongly suggests that their attitudes toward climate change were shaped by conservative media sources, which have been "crucial actors" in publicizing such viewpoints.[69]

The Climate Denial Machine and Evangelical Suspicions of Secular Society

Among the various messaging strategies associated with the climate denial machine, adversarial framing dovetailed particularly well with evangelicals' suspicions of secular society. This was evident, for example, in the claim that climate change activists were employing "scare tactics" in order to achieve some covert goal. Opinions were divided about the motivation for doing so, but the general consensus was that someone (politicians, scientists, or environmental activists) wanted to make people afraid in order to get them to do or support something. The pastor of the Christian Church suggested that scare tactics helped muster support for the solar industry. Similarly, Cam, a man in his fifties who worked as an administrative assistant at a Lutheran church, noted that Al Gore had "invested heavily in companies that will be benefited by the global warming scare."[70]

Beyond suggesting financial motives, however, some informants discerned an anti-Christian motivation. For example, several of the Christian Church focus group participants believed that those who used frightening language when talking about climate change did so in order to encourage people to turn to the government for help instead of to God. Similarly, at the Independent Baptist church, Kenneth argued that

> global warming is an ideology to control you, and me. It's really not anything to do with the environment. It is an ideology, a way of thinking that says, there's a way to live that's better than the way we're living now. . . . I believe it's an idea that's a dangerous idea, that will take us further from God and further from the way America has been in the past.

Here Kenneth interpreted climate activists' warnings about the dangers of climate change as an attack on his faith. Drawing on evangelicalism's long-standing sense of displaced heritage, he further suggested that they did so in order to force Christianity from its former place of cultural dominance. The various ulterior motives that my informants attributed to climate activists demonstrate well how easily secular and religious justifications for skepticism blended.

A second realm in which adversarial framing appeared to resonate especially well among my informants had to do with attitudes toward the government. At the Christian Church, for example, Kelly was highly suspicious of the government's motives. Piggybacking on another participant's comment that climate research was untrustworthy because "you can pretty much make research say anything you want to," Kelly added:

> I believe the government knows that whoever controls the mind of our children controls the next generation. And so there's reason behind everything that's put out there. . . . To me it's fear. I think they try to cause us to fear. And when you fear you want to have someone to trust in.

A number of my informants expressed similar fears about climate-related "scare tactics" being used as a pretext for increased government control. This is no coincidence, for one of the ulterior motives that those associated with the climate denial machine frequently attributed to their opponents was the desire to increase government control, which was presented as threatening individual freedom and economic rights.[71] As we have seen, this argument would be particularly powerful for evangelicals, since they blame many of the societal changes they lament on the federal government's intrusion.[72]

THE EMBATTLED MENTALITY AND SKEPTICISM OF SCIENCE

Social scientific research has shown that evolution skepticism and climate skepticism correlate with each other, and a number of observers of the evangelical climate movement have suggested that evangelicals' distrust of climate science stems in part from the creation-evolution debate.[73] This debate is thought to have imposed "a general culture of scientific skepticism" in the evangelical community, a perception that sociological research has reinforced.[74] Recognizing the role of the embattled mentality in shaping traditionalist evangelicals' environmental attitudes, however, suggests a different interpretation of the association.

What I found was that my informants accepted science in general. However, they distrusted what scientists said about the specific topics of evolution and climate change, in both cases because discussions about evolution and climate change often led to "taking God out of the equation." In other words, it was the embattled mentality, not distrust of science, that was the common thread connecting opposition to evolution with opposition to global warming. As a result, the "culture of scientific skepticism" existed only insofar as it encouraged people to resist specific areas of research that they felt contradicted their understanding of biblical teachings. It did not amount to an across-the-board rejection of science.

That my informants accepted science in general was evident from the focus group discussions. Zach at the Presbyterian church, for example, was suspicious about the scientific research on climate change (he believed that climate scientists were after grant money), but when I asked him whether he felt science in general was trustworthy, he replied, "I'd be a fool to deny that. Science has brought us some great results." Pastor William from the Assembly of God church also argued along similar lines, observing, "You wouldn't have some of our medicines we have today if it wasn't for science. And if you look in the scriptures, Luke was what? A doctor and a scientist. . . . So God uses physicians and scientists." I did not hear anyone argue that science in general was useless or otherwise problematic.

While maintaining positive views about science in general, however, my informants rejected scientific research that conflicted with biblical teachings. In a discussion on the scientific research on climate change, Tyler from the Church of Christ articulated the problem in the following manner: "I think scientific research is important, I really do. [But] without God in the equation it's doomed to fail." Pastor Luke of the Lutheran church objected to evolution on the same grounds, saying that "those who I find run with Darwinism, are trying to take God out of the picture." It is clear from the parallel language that informants used to talk about climate change and evolution that the problem many of them had with both topics had to do with a perceived conflict with biblical teachings.[75] Indeed, at one point my informants made this connection explicitly. According to Peter and his wife, Michelle, the difference between Christian and environmentalist perspectives was this:

> Peter: Christians typically believe that God is in control, whereas somebody that I would label an extremist, an environmental extremist, would say that "I'm in control of what happens, and we've got to save the earth *because God doesn't have the power to*" [emphasis added]. And therefore you unite that with the fact that if they've got the power to do it then they also believe in the whole evolution thing. . . . And so then you start taking God out of the equation completely!

> Michelle: And if I take Him out of that equation I take Him out of other equations. And that's not an option for me. It's not going to be an option in my home.

The lengths to which my informants went to argue that science did not actually conflict with biblical teachings underlined how important it was to them to maintain a general support for science. A common way they did so was by arguing that if conducted properly, scientific research would not conflict with biblical teachings. Pastor Roger at the Independent Baptist church acknowledged the importance of scientific research, including on the issue of climate change, but he clarified that science needed to be what he called "true science":

> I think it's fair to evaluate new scientific things that come as a result of observable scientific-method-type things, but if there's a clear contradiction with the Word of God, then my question would be on the science end of it, rather than on the Bible end of it. . . . True science would be that which agrees with the Bible, but I'm not saying that anything that comes up that even appears to disagree with the Bible is automatically false, I think it should be investigated. Let's test it, let's go through the proper methods. And once we come to a verifiable conclusion, I think we'll probably find that that is complemented in the scripture rather than condemned.

His reference to questioning "the science end of it" rather than "the Bible end of it" made clear where his allegiance lay, while also not putting him in a position of having to deny that science had much to offer. At the same time, one can infer how important his faith was to him from his confidence that any close investigation would reveal that science confirmed scripture.

Another common means of maintaining a general support for the Bible was to accept a limited version of the theory of evolution. Many of the people with whom I spoke told me that they accepted what they called "microevolution" or "adaptation," the idea that species can change over time. They distinguished this from "macroevolution," the idea that one species can give rise to another species. Not coincidentally, the latter is the part of the theory of evolution that they perceived as conflicting with the biblical account of human origins. Accepting microevolution while rejecting macroevolution was thus a way of accepting what could not be denied (given common sense) while denying what could not be accepted (given their faith commitments). It allowed them to coexist with the modern scientific worldview without fully accepting it.[76]

In sum, it was not a general culture of scientific skepticism stemming from debates over evolution that led to climate change skepticism. This argument mistakes the essence of evangelicals' objection to evolution and to climate change. What my informants were concerned about were those aspects of science that threatened their faith, specifically those that they perceived as advancing a secularist worldview. This outlook shaped their attitudes toward both evolution and climate change but, perhaps encouragingly, not necessarily toward other areas of scientific research.

After realizing how the embattled mentality shaped attitudes toward climate change, I returned eagerly to the DVD that Pastor Joshua had given me, wondering what *The Silencing of God* might have to say about the environment.[77]

True to the cover's promise, the film encapsulated much that evangelicals disliked about the direction in which American society was headed. A five-hour-long jeremiad delivered by a single, impassioned speaker, the film described in detail the "sinister forces"—pluralism, atheism, humanism, evolution, liberalism, and political correctness—that were waging an "aggressive assault on the Christian religion." The speaker, a man by the name of David Miller, went on to discuss the Christian beliefs of America's founding fathers and the many ways in which Christianity had historically played a central role in American culture. Yet, he observed, in the last fifty years social and political liberals from the "Left Coast," Hollywood, universities, the American Civil Liberties Union, feminists, "abortionists," homosexuals, and liberal politicians had conspired to reject this heritage. Not one for understatement, he showed a slide listing Hollywood, universities, and liberal politicians under the heading "America's Trilogy of Spiritual Terror."

Four hours in, I had almost given up on the idea of hearing anything about the environment. But Miller turned to it, unexpectedly, in the final hour of his lecture, in a section titled "Solutions." The United States had lost God's favor because it had strayed dangerously far from God's word, Miller reminded his audience. It was not just militant homosexuals or abortionists who were to blame, however, but the Christians who had been too apathetic to stop them. Too often, he admonished, Christians did not vote their values or speak up loudly in public debates: "When you think about the political issues that face our country, upon what basis do you form your opinions?" A lot of people were concerned about the economy or trying to improve the country's educational system, he observed, but what would Jesus think about this kind of concern if he returned tonight?[78] "I would suggest to you that such matters should not be the ultimate and final reason for making our political choices," he intoned. Continuing in the same vein, he asked a rhetorical question that seemed to provoke titters from the audience: "What about the environment? Are you aware of any civilization in human history that was called to account and eradicated because they took the position that the spotted owl needs to be saved?" He smiled at the obviousness of the answer. "No." "What about the snail darter? No. What about any animal in all of human history?" People who worried about animal rights had "lost their spiritual moorings" and strayed from God's view of the world. What they needed was to remember something that Miller and his audience already knew: "This environment will last *just*"—Miller drew out the word for emphasis—"as long as God wants it to last. At which time *He* will destroy it."

After this brief dismissal of environmental concern, he moved on to remind his audience that liberals, feminists, abortionists, and homosexuals were willing to mobilize, organize, and sacrifice time and effort. "I ask you: Are we [Christians] not willing to mobilize, organize, spend time and money? If they do, why can't we? More importantly, why *won't* we?" He ended with a call to action: "May God bless His people in their efforts to represent Him *accurately, forcefully, publicly,* to a civilization that is literally rushing down a precipice to ultimate destruction. *God is waiting for us to stand up and be His mouthpiece.* May God bless us to that end." Despite his frequent reminders that Jesus might return at any moment, his was not a call to otherworldly speculation, but to this-worldly engagement.

CHAPTER 5

How Evangelical Subcultural Identity Sustains Climate Skepticism

To talk about climate change "attitudes" or "opinions" is to conceptualize them as traits that belong to individuals. But as we all know, it is rare for anyone to form an opinion in complete isolation from any sort of community. One of the things that impressed me during my time in the field was the numerous ways in which my informants' climate change skepticism was woven into the social fabric of their lives. Indeed, this became all the more apparent because I could see the many ways in which *acceptance* was woven into mine.

Despite its evident importance, the role of sociality has been underexplored in research on Christians' environmental attitudes. The reasons are probably historical: as mentioned in the introduction, the theoretical paradigm for most of the social scientific research on Christians' attitudes toward the environment has been Lynn White Jr.'s argument that Christianity discourages environmental concern because of its anthropocentric vision of creation.[1] The dominant influence of the "Lynn White thesis," as it is called, has resulted in an almost exclusive focus on how Christian *ideas* might shape attitudes toward the environment, whether they are ideas about human dominion, the dangers of paganism, the overriding importance of salvation, or the imminence of the end times.

From my perspective in the field, it was evident that my informants' environmental attitudes were indeed connected to theological positions. But at the same time, they were in many ways *sustained* by social practices. In the first place, which issues we pay attention to and which we ignore have a lot to do with those around us. Hence, my informants' lack of interest in or awareness of environmental problems (as discussed in chapter 2) and comparatively intense interest in examples of Christian persecution (as discussed in chapter 4) constitute one powerful way in which their social world was acting to shape their environmental attitudes. Second, it was also clear that my informants' attitudes toward the environment were not just about their attitudes toward the natural world but also about their attitudes toward environmentalists—that is, about their attitudes toward a specific group of people. Environmentalists were in fact socially useful for my informants because environmentalists served as an out-group against which they could favorably contrast themselves. In this sense, my informants' environmental attitudes were sustained by another social practice, that of identity work. Finally, comparing my evangelical informants' attitudes toward the environment with those of my Seventh-day Adventist informants underlines the extent to which these particular social practices were tied to evangelical subculture. That is, my informants' attitudes toward the environment were embedded not only in their social world, but also in social practices whose intelligibility depended on a broader narrative that many traditionalist evangelicals tell themselves about the rightful place of Christianity in American society.

THE SOCIAL NATURE OF ATTENTION
TO ENVIRONMENTAL ISSUES

As the cognitive sociologist Eviatar Zerubavel pointed out in his book *Social Mindscapes* (1997), the process of learning what to notice and what to ignore in the world around us is fundamentally social. We learn what is relevant and irrelevant from those around us, and we internalize these choices so thoroughly that relevance seems to be intrinsic to the phenomenal world, rather than what it really is: a result (largely) of choices that we are taught to make. As Zerubavel points out, such choices are learned through participation in different "thought communities"—groups such as churches, professions, political movements, or generations—that shape our habits of attention and concern. Thought communities teach us to see the world through a particular lens, one that highlights certain features as salient while de-emphasizing others as unimportant background information. These communities have wide-ranging power when it comes to our social lives, for they teach us to "assign to objects the same meaning that they have for others around us, to both ignore and remember the same things that they do, and to laugh at the same things they find funny."[2]

In the case of my informants, it was clear that their churches and the broader evangelical subculture of which they were a part functioned as thought communities, directing their attention and concern toward certain topics and away from others. Of course, the most powerful and pervasive evidence of this was the fact that certain issues came up repeatedly at different churches—prayer in school, abortion, declining morality, and the perceived marginalization of Christianity—while there was little consistency in the environmental problems they mentioned, if they mentioned any at all. While this contrast is not commonly mentioned in discussions of Christians' environmental attitudes, it struck me as one of the most powerful barriers between the Christians I met in Georgia and the environmentalists of my acquaintance. In a cognitive sense, they lived in different worlds.

If thought communities establish our mental horizons, determining which issues are worthy of notice, they also establish our *moral* horizons, determining which issues are worthy of moral reflection.[3] Among my informants, it was clear that the environment was generally not one of these issues. While the evangelical thought community brought morality into focus, distinguishing certain moral issues from the "background" of other issues to which they could possibly attend, the focus was on a narrow set of issues, most of them related to sexuality. Since the number of issues anyone can pay attention to is limited, this cast environmental problems (along with many others) into the background. The link between attention and moral concern is particularly clear here, for the fact that environmental problems were mostly invisible to them (as described in chapter 2) ensured that they were morally invisible as well. Environmental problems were behind a curtain, generally excluded from moral contemplation because they were largely excluded from notice in the first place.

When I lifted the curtain, so to speak, explicitly asking my informants whether they considered harming the environment a sin, all the pastors I interviewed and most of the focus group participants agreed that they should not harm the environment, but many rejected the notion that it would be sinful to do so. As one pastor told me, "We've been told to be good stewards. To say it's a sin, I can't do that." In the focus groups, whether environmental harm constituted sin was a matter of debate. Some people were willing to describe purposeful harm as sinful, but others countered that it should be placed "low on the list of priorities" or removed entirely, since sin was a "transgression of God's law" and harming the environment would not necessarily constitute such a transgression. Given the central role of sin for evangelicals—sin not only denotes "*wrong* in the deepest sense of the word,"[4] but explains all manner of earthly imperfections—the tendency to exclude the environment greatly weakened the case for environmental protection as a moral responsibility.

When it came to climate change, the lack of clear scriptural guidance on the issue seemed to encourage my informants to be especially attentive and deferential to cues from their thought communities.[5] For instance, Ben, who pastored an independent Church of God, explained to me that there was a clear Christian stance on cigarettes and pornography but not on climate change because

> those issues have been handed down from previous generations. . . . So you've got all of church history as well as the last hundred years of preaching material to ascertain truth from, whereas [climate change] is kind of, on my end, and maybe it's not in its infancy, but as far as my knowledge, it's elementary.

Ben's candid reference to the importance of tradition seemed to highlight that determining the appropriate stance toward a particular moral issue was not simply a matter of biblical exegesis but entailed achieving social agreement. That is, determining biblical morality was in part a social endeavor. His comment about having access to "preaching material to ascertain truth from" was particularly revealing in this regard: social positions did not simply emerge from scripture but were developed slowly, ideally (from his perspective) with the confidence that others' interpretations provided. Exchanges such as these underlined how my informants' attitudes toward climate change were shaped by their churches as thought communities.

ANTI-ENVIRONMENTALIST ATTITUDES AS IDENTITY WORK

The second way in which it was clear that my informants' attitudes toward the environment were shaped by social practices had to do with their repeated identification of environmentalists as an out-group. Identifying out-groups is useful because it helps create a sense of collective identity: clarifying who the group is *not* also helps explain who it *is*.[6] For my informants, environmentalists were "radicals" or "crazy people" who were obsessed with things that did not ultimately matter, and spiritually suspect in that they likely harbored pantheistic tendencies. By differentiating themselves from environmentalists, my informants helped establish a positive identity for themselves as moderate, sane, or sensible people who cared about things that really mattered and remained true to their faith in doing so. The process of collective identity formation and maintenance, necessary for social mobilization, thus created friction—or, from another perspective, reinforced social boundaries—between evangelicals and environmentalists.

"I Am Not a Tree Hugger": Mapping Evangelical Identity

One way to understand how groups construct identity by means of contrast with other groups is through the concept of "mapping."[7] The term *mapping* is useful because it highlights the geographical qualities of identity work, in which groups typically locate other groups as socially "near" or "far" from themselves. At the same time, like maps, identity work is also descriptive, revealing the social territory that "belongs" to the group.[8] When it came to the environment, my informants spent a significant amount of time "mapping" the boundaries of their identity in ways that excluded environmentalists. For example, as discussed in chapter 2, when I asked what environmentalism meant to them, many responded that environmentalists were radicals or extremists who "went too far." By implicit contrast, they established their own identity as sensible moderates. When Renee at the Southern Baptist church described her openness to the idea that climate change might be occurring, for example, she made a point of rejecting the notion that this made her an environmentalist, which she equated to being a radical:

> I'm out to lunch on [whether climate change is occurring]. I'm not sure, because there have been periods of time where the earth has warmed and then cooled. But I think in light of the fact that we have progressed to this state where we have planes and trains that have never been on the earth, it makes sense that it would have warmed the earth. . . . I'm not saying that it—*I am not a tree hugger, one of these radical people,* but I just don't see how the things that are happening today, the technology, could not have affected the atmosphere. (emphasis added)

While wanting to entertain the idea that climate change was occurring, Renee clearly struggled with the perception that this might lead others to view her negatively. In drawing a line between herself and the environmental other, she attempted to ensure that her opinions would be taken seriously. That she felt the need to do so exemplifies the importance of boundary marking in evangelical circles.

At the fundamentalist Independent Baptist church, when I asked participants how they would define the word *environmentalism,* their responses included "extremists," "radicals," and "crazy people." Michelle even differentiated explicitly between caring for the environment and being an environmentalist, saying that "people that I run in circles with, you would never say, 'I am an environmentalist.' But I would say, 'I think you should take care of the land that you live in.' Because we think if you say 'environmentalist,' you know, [you're saying] 'I'm a crazy person.'" Here she stressed the importance of distancing oneself from an identity that those in her faith community viewed negatively. Identity was a key issue for her, as underlined by her acknowledgment that certain behaviors would be acceptable as long as they were not labeled "environmentalism." In other words, in this case it was not so much behavior but the identity label associated with that behavior that mattered. Pastor Richard at the Southern Baptist church made exactly this point, musing to me at the end of our interview that "perhaps the reason why you haven't heard these clear messages from the pulpit about [environmental] stewardship is because maybe there's our own political correctness and you don't want to look like some *tree hugger* to the rest of the evangelicals" (emphasis added).

Perhaps the ultimate sign that my informants viewed environmentalism as conflicting with their identity as Christians was that a number of them proposed adopting a different term to refer to a Christian who cared for the environment. For example, Ethan from the Presbyterian church clearly had pro-environmental leanings, yet he had learned to be careful how he expressed them: "I majored in environmental science. I don't use the term *environmentalist* anymore because people think you can't associate environmentalism with Christianity. So I've adopted the word *conservationist* for myself just to avoid the extremes that people associate with the term." Ethan's remark underlined the extent to which the evangelicals around him viewed environmentalism as an identity that was incompatible with their faith.[9]

Such comments make it clear that what might have seemed from afar to be simply attitudes about the environment were something more: attitudes about environmentalists. To recognize this is to see environmental attitudes not simply as free-floating ideas that individuals may or may not come to hold, but (for my informants at least) as a kind of identity work. And as my informants' comments made clear, it was *work,* for it required effort to express opinions about the environment while staying "on the map" of evangelical identity.

Faith as an Identity Marker

Not surprisingly, my informants also saw their faith as a key point of contrast between themselves and environmentalists. Informants in five of the nine focus groups argued that the problem with environmentalism was that it promoted a different form of worship. In particular, many informants drew a line between being concerned about the environment (which was acceptable) and worshipping it (which "went too far"). When I asked the group at the Christian Church what constituted crossing the line from concern to worship, Kim, a college student in her twenties, replied that, as a Christian,

> you are supposed to be all about God. Like everything about you, you can do for God. People who are extremists wear clothes that are environmentally friendly, they wash their hair with shampoo that's environmentally friendly. Everything they buy, everything they do is about that. And we're supposed to try to make everything we do about God, not about the earth.

Here Kim pointed to orientation (toward God or toward the earth) as a key boundary marker. In referencing clothes and spending habits, she also highlighted the material trappings of identity. Identity is expressed not only in terms of values, but through visible markers like clothing and more subtle signs of belonging, such as choosing to consume products that reflect one's values.[10] These visible markers are one of the most powerful ways of marking identity in modern life, and clearly one to which my informants were attuned. Pointing out visible differences was thus another way of marking environmentalists as "others"; they were visibly part of another tribe.

The geographical metaphor of "going too far" was particularly apt given that my informants used it when mapping the boundaries of their identity. Krista, an energetic woman in her forties who participated in my focus group at a Christian Church, argued that "you can be a very good responsible citizen and a Christian and take care of what God has given us without going overboard and worshiping it, worshipping the creation rather than the creator." Similarly, Shirley at the Southern Baptist church argued that environmentalism was

> not a bad thing unless they take it to an extreme, meaning it's almost a sense of worship to mother earth. That is an extreme radicalist [sic] or an extreme environmentalist. . . . As Christians we should all be environmentalists, it's our job to take care of the earth, what God has given us, but not to go to the other end of the spectrum per se to [see it as] a god, lower case g-o-d, a god of the earth or a god of nature. That is an extreme.

Krista and Shirley both emphasized the importance of taking care of the environment in the same breath that they cautioned against taking it too far by worshipping it. What marks such speech as identity work is their focus on differentiating proper Christian behavior from the bad/negative/unacceptable/radical behavior of environmentalists. They are not only talking about how to take care of the environment but also about how to avoid becoming the "other" while doing so—and this was typical of how my evangelical informants answered my questions about how to be an environmentalist.

It is important to emphasize that instead of criticizing environmentalists for their radicalism, my informants could have recognized that there are many different kinds of environmentalists, some of whom are more "radical" than others. Instead, however, they tended to associate all environmentalists with extremism, going too far, and radicalism. Doing so was socially useful because it heightened the contrast between themselves and environmentalists, and did so in a way that cast themselves in a positive light.

As Christian Smith and his colleagues have shown, the sense of strong boundaries with the non-evangelical world is a defining feature of the evangelical worldview. It was clear that this desire to maintain strong boundaries affected my informants' attitudes toward environmentalism, which they saw as coming from "the world."[11] For example, Chris, the young pastor of a Presbyterian church, cautioned that

> we [Christians] need to be careful that what society tells us is ethical care for the environment is *actually* ethical care for the environment. . . . I think people can get caught up with what someone else—the media, this social group, that social group—is saying is an ethical obligation, [when] they may be binding your conscience sinfully.

The thrust of his argument—that Christians should not just go along with what society says—is characteristic of evangelicals' attitudes toward a host of issues. For example, a Baptist man who participated in Smith and colleagues' study stated that "there are a lot of pressures that we Christians feel in society because of the values the world espouses today. Look at the pressure to go to the movies that are riddled with sex and unbearable language—that creates real problems and pressures for Christians."[12] For Chris, environmentalism might be one of those values pushed by the secular world. Thus, the sense of being outsiders separate from the secular world provided an additional reason to resist environmentalism: it was a popular concern embraced by those in the mainstream. Remaining skeptical about environmental issues was thus part of a broader pattern of maintaining distinction from "the world."

Scholars have generally regarded evangelicals' fear that environmentalism will lead to pantheism as a theological barrier to environmental concern. It certainly is a theological barrier, but it is more than that: my informants' theologically rooted opposition to pantheism served as a boundary marker between themselves and environmentalists.[13] And not only did theology and identity maintenance work in tandem to curtail environmental concern, but the very differences that my informants highlighted to contrast themselves with environmentalists tied back to their sense of enmity toward secular culture. In its otherness, environmentalism helped define what was wrong with the current state of affairs and provided an opening to highlight what Christianity offered instead.

Humor as a Boundary Marker

A final way in which my informants differentiated themselves from environmentalists was via humor. Humor is an important tool for constructing identity, for it helps define who is in (those who get the joke) and who is out.[14] When I would ask people I met while living in Georgia if they ever heard the topic of global warming come up in conversation, they generally said no. But intriguingly, a few people answered that when it did come up, it was in a joking context. One of my neighbors, a retired teacher, told me that she didn't hear people talk about climate change frequently, but when they did, it was as a humorous explanation for hot summers. In the focus group I conducted at the Christian Church, I encountered this tendency firsthand. At one point, in response to a question about how they viewed the future, Anne responded, "Do I have a 401K? Yes. Do I hope to have some time [on earth]? Yes. But . . . my future, hopefully, if I stay on track, then I will be in heaven, that's my future. Do I have a wedding [coming up]? Yes. *If* we're here." Seizing the comedic moment, Krista, the woman from the Christian Church mentioned above, said, "Lord willing and the crick don't rise." That comment drew a few chuckles, but a moment later the group erupted into uproarious laughter when she added, "Due to global warming!" Since only skeptics would be so unconcerned about climate change that they could treat it as a joke, the comment created a sense of camaraderie by playing up a commonly held belief. At the same time, the joke implicitly defined who was in the group (those who found it funny) and who was out (those who did not)—while also subtly making the point that anyone from the latter group might be subject to mockery. Thus, humor served to reinforce boundaries with environmentalists by implying that accepting climate change was something other (unreasonable) people did, and by demarcating climate change believers as outsiders who could not see the humor in their own misguided assumptions.

In sum, mapping the evangelical identity so as to exclude environmentalists allowed my informants to establish who they were through contrast. In their view, they were a moderate and sensible group, one that was devoted to the struggle to preserve America's Christian heritage, while environmentalists were radicals, extremists, tree huggers, and crazy people who, like Hollywood elites, were causing moral decay by promoting secular values. This sense of enmity toward environmentalists subtly but powerfully discouraged concern about the environment and climate change, for to express such concern was to risk one's good standing in the community and, indeed, to risk one's "insider" status altogether.[15] The embattled mentality was deeply implicated here, for the sense of embattlement with secular culture both fostered the sense of a collective goal and cultural identity for evangelicals and identified secularists (including environmentalists) as barriers to the achievement of this goal. Their relative lack of environmental concern was thus a product not just of theology, but of sociology—a sociology deeply informed by their unique reading of history and attendant sense of identity.

COMPARING EVANGELICALS WITH SEVENTH-DAY ADVENTISTS

The time I spent at a Seventh-day Adventist church in the region further underlined how powerful the broader evangelical culture of embattlement was in shaping attitudes toward the environment. While Adventists share certain features with evangelicals (they have high regard for the authority of scripture, for example, and place great emphasis on spreading the gospel), they do not participate in broader evangelical discourses and are not members of the NAE. Thus, while they are sometimes grouped with evangelicals for the purposes of analyzing political behavior, they are not evangelicals in a cultural sense, and those I met at the church outside of Dixon did not share evangelicals' embattled mentality.[16] Through my focus group and interviews with Adventists, it became clear that they were unlike evangelicals in another way: they viewed environmentalism positively. This nonconfrontational attitude was visible not only at the church I visited but at the denominational level, where Adventist leaders have not only publicly embraced environmental values, but have also utilized educational campaigns to encourage their members to adopt them. I suggest that they may be doing so as part of an effort to fit in with mainstream society. By contrast, many evangelicals eschew environmentalism in part for the very reason that Adventist leaders embrace it: because it represents assimilating mainstream society's values.

Like evangelicals, Seventh-day Adventists are theologically conservative. Adventists believe, for example, that the Bible is the inspired word of God and the "infallible revelation of His will."[17] In addition, Adventists also strongly emphasize the end times—there was a plaque declaring the nearness of the end times in the front entryway of the church I visited—so any differences in environmental views could not be attributed to eschatology. Yet there were also important differences between the Adventists and the evangelical churches I visited, differences that were immediately apparent to me. First, the composition of the Adventist church was quite racially diverse. On the days I attended services, I estimated that about 20 percent of the congregation was African American, and there were other ethnicities present as well. This contrasted sharply with evangelical churches I had visited, most of which were almost uniformly white (except the Church of Christ and Assembly of God churches, which were slightly more diverse). Another difference was a more international membership. Not only did I hear attendees speaking with foreign accents, but the Adventist pastor himself was not American (though he was from an anglophone country). By contrast, all the evangelical pastors I interviewed were American, if not necessarily born and raised in the South. Finally, and most importantly, when we spoke, there was a difference in the Adventist pastor's tone that was immediately noticeable. When I asked Pastor Matthew of the Adventist church if Adventists believed the Bible was inerrant or infallible, he responded that

> we believe that the Bible is the inspired word of God. We do not have the inerrancy of scripture—*we don't have that issue in our church*. We believe that God communicated His will and His plan of salvation through people, and obviously we take very seriously God's word. *But we don't basically fight over that issue*. (emphasis added)

Like their pastor, the Adventists who participated in my focus group also seemed to have a less combative attitude toward society. Rather than the marginalization of Christianity, for example, Adventists were more concerned about challenges related to their specific denomination. In the focus group discussion, for example, a young woman mentioned that it was difficult for her friends to understand why she didn't drink, and another woman who was a teacher mentioned that she had to explain why she couldn't attend the school's football games, which were held on the Adventist Sabbath (which begins at sundown on Fridays). In short, the Adventists I met were concerned about how to get along with non-Adventists around them rather than about Christianity being under attack.

The Adventists I encountered also displayed a less adversarial attitude toward environmentalism, a difference that manifested itself in several ways. First, they had a positive view of what it meant to be an environmentalist. When I asked the focus group what the term *environmentalist* called to mind, instead of offering responses such as the ones I heard at the evangelical churches, James offered a much milder description: "For me it's just someone that cares more. I think of them as more of a normal person that just recycles more than the average person, or drives a Prius [laughter from the group]. I drive those too, so. But just someone who's a little more aware." Notably, even though the rest of the group laughed when he mentioned driving a Prius, James hastened to make sure his comment did not come off as a criticism by adding that he could also be placed in that category. His description of environmentalists as "someone who's a little more aware" did not imply that they were cultural opponents, but that they were part of a continuum in terms of the seriousness with which environmental problems could be regarded.

My Adventist informants also saw a link between environmentalism and their tradition's focus on health. Due in part to the efforts of the early Adventist leader Ellen G. White, Adventists have long advocated healthful practices such as vegetarianism and avoidance of alcohol and tobacco. The positive effects of these practices have been verified in the medical literature and popularized through books such as *The Blue Zones: Lessons for Living Longer from the People Who've Lived the Longest,* which my informants proudly suggested I read.[18] Because of this history, while my informants acknowledged that the environment was not a common sermon topic, they nevertheless saw it as a concern that connected with their denomination's historical commitment to health. When I asked the group whether they felt they had any obligation to care for the environment, Evelyn, a middle-aged white woman of South African origin, responded that if we believe that God created the world then we need to take care of creation. And that's the whole reason really that we try to keep Sabbath. It's like a cycle, and the seven days, we're basically commemorating creation, worshipping our creator. So I think we need to be examples to society taking good care of our environment.

Her reference to being a good example to society further suggested that she, as an Adventist, felt (and sought) a sense of solidarity with society, rather than opposition to it. Charles, a retired professor, drew a similar connection between Adventists' concern for health and their concern for the environment: "I can honestly say I've never heard an Adventist preacher preach on the environment. We don't simply focus on those kind of things. . . . But it's in part of the culture. If you care for the body you have to." Like these Adventists, the evangelicals I spoke with also affirmed that they had an obligation to be good stewards, but the evangelicals were not able to normalize such concern as a logical expression of their tradition's values in the same way that the Adventists could.

What these differences point to is that evangelicals and Adventists belong to different subcultures, and these subcultures have oriented each group differently toward society and, by extension, toward environmentalism and climate change. Whereas evangelicals tend to be combative, Adventists have increasingly attempted to harmonize their views with those of mainstream society. One explanation for this difference is that Seventh-day Adventism is a sect in the process of becoming a denomination. As the sociologists Rodney Stark and William Sims Bainbridge explain, sects are religious groups that exist in a state of high tension with the surrounding culture, while denominations exist in low tension.[19] Sect or denomination status is not immutable, however. According to church-sect theory, over time sects tend to become denominations because as the members of sects gain social respectability, they seek to downplay tensions in order to better assimilate. Ultimately this process tends to turn sects into denominations.[20]

Adventism began as an apocalyptic sect, but as its members prospered and became more integrated into society, its apocalyptic doctrines became less convenient and its seminaries and leaders consequently began to downplay them.[21] Apocalypticism is still an important recruitment tool—which explains why I encountered apocalyptic views among some of the focus group participants—but at the leadership level, Adventists are attempting to reduce tensions with society.[22] At the same time, those I met were also more willing to embrace environmentalism than were the evangelicals I contacted. If one accepts church-sect theory, this is no coincidence, for accepting environmentalism is one way in which Adventists have embraced the societal mainstream.

While the Adventists who participated in the focus group were not necessarily representative, their responses do not seem anomalous when put in the context of official Adventist teachings about the environment. Among their official positions on current issues, the denomination includes a statement about "Stewardship of the Environment." It is strongly worded, citing "massive" rainforest destruction, the hole in the ozone layer, and "dire predictions of global warming, rising sea levels, increasing frequency of storms and destructive floods, and devastating desertification and droughts."[23] This is quite a strong statement on climate change, given that it was issued in 1996, when the science supporting climate change was less definitive. It is also interesting to note that, like the Adventists who participated in my focus group, the statement links environmental stewardship with the tradition's focus on health, concluding that "Seventh-day Adventism advocates a simple, wholesome lifestyle, where people do not step on the treadmill of unbridled over-consumption, accumulation of goods, and production of waste."[24]

In addition, the denomination had released a statement in 1995 that specifically addressed "the dangers of climate change."[25] The statement recommended various actions at the governmental level, further indicating their acceptance of conventional political mechanisms (and contrasting with evangelicals' general distaste for governmental intervention). This does not mean that all Adventists embrace environmentalism, but their leadership has clearly embraced environmental concern.

Adventist leaders have also gone beyond issuing statements. This says something about the strength of the leadership's interest in environmental issues, for many denominations never move beyond the initial step of issuing a statement.[26] In their case, Adventists have incorporated the "creation care" message into their Bible study materials, which are distributed worldwide. I became aware of one such effort through a chance encounter with an energetic man named Kent who taught drama at the school affiliated with the church I visited. As I was chatting with other church members before the focus group, Kent, who had mocha-colored skin and striking sea-green eyes, stopped by to express his interest in its environmental theme. Talking in his classroom a few days later—a room filled with the gauzy remnants of performances past—he professed a long-standing fascination with the ocean. With evident pride, he handed me a news article from 1996 that described his work with an environmental nonprofit he had headed in a Caribbean country before coming to the United States. In the article, he is quoted as criticizing the "tremendous apathy" people have about littering and wondering "what the children will have to inherit, this paradise which we enjoy now or a garbage dump?" Although he told me that he accepted Adventist teachings on the end times, he also emphasized in our conversation that it was still incumbent upon Adventists to take an active role in saving and maintaining the environment—and he had made good on his word.

Knowing of my interests, Kent also handed me a copy of the church's *Adult Sabbath School Bible Study Guide* for January–March 2012. This pamphlet was prepared by the Office of the Adult Bible Study Guide of the General Conference of Seventh-day Adventists, which "reflects the input of worldwide evaluation committees and the approval of the Sabbath School Publications Board." Hence it presumably had global support, at least at the leadership level. The reason Kent had set it aside for me was that one of the weekly lessons was on "Creation Care." As it turns out, the lesson's primary subject is the tension between caring for the environment and expecting Jesus's imminent return. While not eschewing the denomination's apocalyptic teachings outright, the study guide emphasizes the need to care for creation and, in the process, urges readers to balance apocalyptic expectations with environmental concern.

The study guide provides strong evidence that Adventists have consciously, publicly embraced environmentalism in a way that it is difficult to imagine my evangelical informants doing. While the study guide eschews radicalism when it comes to the environment, much as my evangelical informants had, it does so without employing combative language. This further suggests that the combative worldview is a key factor that keeps evangelicals from embracing environmentalism.

Examining the study guide in detail underlines this point. Divided into seven lessons, one for each day of the week, it opens with the following question for contemplation on Sabbath afternoon: "What should we, as Seventh-day Adventists, think about the environment, especially because we know that this earth is corrupted, will continue to be corrupted, and will one day be destroyed?" Since humans are commanded to "have dominion over the fish of the sea, and over the fowl of the air, and over the cattle, and over all the earth, and over every creeping thing that creepeth upon the earth" (Gen. 1:26), the authors concede, it is "no wonder that, at times, we struggle with how to relate to environmental concerns."[27]

Throughout the daily lessons, the study guide recommends what might be described as a moderate version of environmental concern. Sunday's lesson, the first of the week, urges Adventists to avoid "extremism," depicted via a tale about environmentalists who (foolishly, it is implied) devoted themselves to liberating live lobsters from fancy restaurants. The authors point out that the promise of a new heaven and a new earth does not give Adventists license to exploit the present earth, but also that they should be careful not to make an idol of the earth. Notably, while the study guide pokes fun at the "lobster liberation movement," it does so in gentle language that does not denigrate environmentalists or otherwise position them as an out-group ("Caring about the environment is one thing, but stealing a lobster out of a restaurant and taking it, by helicopter, back to the ocean does seem a bit extreme, does it not?").[28] The next day's lesson reprints the church's official statement on the "Environment" while also admonishing Adventists to care for the environment because "as Christians who believe that this world and the life and resources on it are gifts from God, we should be at the forefront of seeking to take care of it."[29]

For Tuesday's lesson, the study guide asks readers to contemplate the environmental implications of the biblical injunction to "love your neighbor as yourself."[30] Returning to Genesis, it also notes that Adam was "called to be a steward. . . . God didn't tell him to exploit [the earth]. . . . Instead, [Adam] is told to work it and protect it." On Wednesday it asks readers to contemplate how the Sabbath might be relevant to the environment, especially in "commanding us to rest from our work, to rest from seeking to make money and do business." On Thursday it asks readers to contemplate Genesis 1:26 and to remember "humankind's responsibility, as ruler of the world, to take care of it, because God created it, and it was 'very good.'" On Friday, it summarizes the week's lessons by envisioning a balance between the knowledge that Jesus will be returning soon and the responsibility to be good stewards of the earth's resources:

> No question, this world is coming to an end; it will not last forever. And yes, Jesus is coming soon. All that's true, but nothing in these truths gives us the right, or the mandate, to defile the earth. If anything, as Christians, we should seek to take care of the world that our God has created for us.[31]

In all, the study guide encourages a moderate version of environmental concern, one that recognizes humanity's impact on the environment and its responsibility toward it while avoiding engaging in "extremism" or idolatry. Importantly, it makes a concerted effort to counter the apathy-inducing consequences of end-time belief, while not denying (but not emphasizing either) that the end is coming. Underlining the importance of caring about the environment even if the Second Coming is near, on the last day the study guide asks readers to discuss how they would respond to the question "Jesus is coming soon, so why should I care about the environment?"[32] Kent told me that when his group discussed this question they had no trouble coming up with answers.

While the Adventist leadership's preference for moderation was therefore similar to that of the evangelicals with whom I spoke, it was expressed in a different tone that did not describe environmentalists as outsiders or crazy people. Moreover, some of the Adventists I met had taken an active interest in environmental issues, actions that had little counterpart in the evangelical churches I visited. This is not to suggest that all Adventists are strong environmental advocates or that some Adventists have not perhaps absorbed from their evangelical brethren an antipathy toward environmentalists, especially since evangelicals have so powerful a voice in American culture that few are out of ear-shot. Indeed, there is little evidence that Adventists have made great headway in moving from concern to activism at the denominational level. But the Adventists I met had come much further than had the evangelical churches I visited. If Lawson is correct that Adventist leaders are trying to gain mainstream acceptance, their promotion of environmental concern makes sense. Conversely, since many evangelicals seek the opposite—to differentiate themselves from mainstream society—it also makes sense that they would reject or be skeptical of environmentalism.

One might argue that leaders in the evangelical tradition are not so different from leaders in the Seventh-day Adventist tradition in terms of acknowledging the importance of the environment. The NAE has, in fact, repeatedly expressed its support for environmental stewardship, beginning with a resolution passed on "Ecology" in 1970.[33] In 2015 it even passed a resolution on "Caring for God's Creation: A Call to Action," which acknowledged that "a changing climate threatens the lives and livelihoods of the world's poorest citizens."[34] But more relevant to this discussion than the leaders of the NAE are the evangelical leaders who have helped nurture a sense of tension with and distinction from mainstream society. As Part Two of this book will show, it is the latter group that has been most active in promoting climate skepticism. The NAE's position, by contrast, appears to be out of step with the views that I encountered in Georgia and with the rising levels of climate skepticism among American evangelicals that surveys have revealed.

In sum, my comparison of lay Adventists with lay evangelicals in Georgia supports the argument that theology is only part of the story when it comes to understanding how a religious tradition shapes its adherents' environmental attitudes. The contrast between the two groups' attitudes toward the environment suggests that the broader religious subcultures in which each participated were key factors in shaping those attitudes. For Adventists, this meant an attitude of acceptance and accommodation of environmental values at the leadership level—and, from what I saw, among laypeople as well. By contrast, the combative attitude that permeated evangelical subculture also fostered suspicions about environmentalism and environmentalists, whom they viewed as cultural opponents.

My encounter with the Seventh-day Adventists marked a turning point in my research, suggesting that what played a pivotal role in shaping my evangelical informants' environmental attitudes was not simply individual-level opinions about politics, theology, or science, but the social practices embedded in evangelical subculture itself. Given this finding, I was surprised and intrigued to learn from one of the pastors I interviewed that there had recently been an initiative within the Southern Baptist Convention (SBC) to raise concern about climate change. The SBC is well known for its religious and political conservatism, so I wondered how and why, in this case, concern about climate change had been able to emerge.[35] What did it take to overcome the powerful subcultural barriers to concern about climate change that I had observed among my informants in Georgia? The recent upsurge of concern within the SBC offered a chance to better understand the embattled mentality's impact.

CHAPTER 6

Salt and Light

Skeptical Environmental Stewards
of the Southern Baptist Convention

The SBC is the largest Protestant denomination in the country. With its reputation for conservatism, it probably surprised few observers when in 2007 it approved a resolution on global warming that urged Southern Baptists to "proceed cautiously in the human-induced global warming debate."[1] The resolution repeated many of the skeptical arguments current at the time—the scientific community was divided about the reality of global warming, global temperatures have risen and fallen cyclically throughout history—while also making specific objections to certain policy options, such as signing on to the Kyoto Protocol. "Exempting emerging economies like China, India, and Brazil from CO_2 and other greenhouse gas emissions reductions," it announced, "would significantly undermine the minute effect on average global temperature gained through reductions by developed nations."[2]

Just a year later, however, forty-six influential Southern Baptists released a statement that appeared to indicate a reversal.[3] According to a story in the *New York Times*, the statement "call[ed] for more action on climate change, saying [the denomination's] previous position on the issue was 'too timid.'"[4] What was most significant about "A Southern Baptist Declaration on the Environment and Climate Change" (hereafter "Declaration"), according to the *New York Times* and other news outlets, was that it seemed to demonstrate that Southern Baptists were not a monolithic group of climate skeptics. Instead, the Declaration appeared to indicate that a small but powerful group of Southern Baptists now believed it was important to take action to stop climate change. Taken together with the Evangelical Climate Initiative (ECI), the publication of the Declaration suggested that evangelicals appeared to be on the verge of a historic shift away from the issues that had dominated their political agenda over the past thirty years.

To understand why the evangelical subculture was not, apparently, having the same effect of discouraging concern about climate change among Southern Baptists as it had among my informants, I decided to interview a group of pastors who had signed the Declaration from the general region of my fieldwork, including Georgia, northern Florida, and South Carolina (appendix A describes how I identified these pastors and the dates of interviews). The news stories I read about the Declaration had described it as announcing support for action on climate change, so I thought I was going to be interviewing theologically conservative pastors who had become convinced of the need for action. Instead, to my surprise, all but one of the eleven signatories I had chosen to interview were climate skeptics. In subsequently analyzing news coverage of the Declaration, I realized that many of its prominent signatories (with the exception of its creator) had expressed climate skepticism in interviews with the press. Some of these leaders even argued that the Declaration was compatible with the SBC's officially skeptical stance.

How could a group of climate skeptics possibly have believed that a declaration that supported action on climate change was compatible with their views? Puzzled, I reread the Declaration carefully. Did the document advocate action to halt climate change or did it express climate skepticism? Incredibly, I realized that the wording was ambiguous enough that it could be read either way—depending on what interpretive lens the reader used. Because the text had elements that called different frames to mind for different audiences, two different (indeed, almost opposite) interpretations of the Declaration's position on climate change emerged.

The national news media and scholars read the Declaration through the lens of a narrative that had developed by the mid-2000s about what was occurring within the evangelical tradition. According to this "greening of evangelicalism" narrative, the tradition was slowly shifting toward greater environmental concern, especially by tackling the issue of climate change.[5] Using this lens, and adopting what I call the "Climate Change Interpretation," this group understood the Declaration to be calling for action to halt anthropogenic climate change. By contrast, the Southern Baptist pastors I interviewed and other leaders who spoke up in the press read the Declaration through the lens of their sense of embattlement with secular society. From their point of view, which I call the "Stewardship Interpretation," what the Declaration called for was moderate environmental concern that did not necessarily entail accepting that climate change was a real and serious problem caused by human activities. Confusingly, they described this position as environmental stewardship.

While there is no single correct interpretation, in the end the Stewardship Interpretation mattered more, because it was the one that indicated in which direction the tide was turning.[6] While journalistic and scholarly accounts from this period hailed the Declaration as further evidence of a coming sea change within evangelicalism, examining how Southern Baptist signatories were interpreting it yielded a more prescient interpretation: harbinger of a powerful backlash against climate change activism within evangelical circles. Comparing the two interpretations also puts into sharp relief how the embattled mentality shaped signatories' views about climate change: rather than desiring to ally with climate activists, what they wanted was to articulate a distinctly Southern Baptist position, one that would contrast with the message that secular environmentalists were offering. In taking a stand on climate change, that is, they did not intend to announce solidarity with environmentalists, but rather to wrest the environmental narrative away from (as the president of an important Southern Baptist seminary put it) pantheists and liberal environmentalists.

This is not to suggest that the Declaration's many signatories were all climate skeptics who were suspicious of the environmental movement. But neither did the Declaration represent unified voices in support of action on climate change, as the secular press portrayed it. Instead, the Declaration represented a moment of heightened tension and debate about what, if anything, the evangelical contribution to the climate debate should be.

CREATING THE DECLARATION

The Declaration's origins can be traced to Jonathan Merritt, who conceived of the idea while a student at the SBC-affiliated Southeastern Baptist Theological Seminary (SEBTS). Merritt described the experience as an epiphany:

> I was sitting in class and we were talking about the revelation of God and Dr. Hammett made a statement that was striking. He said and the book recounts that God speaks to us both through general revelation in nature and the

special revelation in the scriptures, and when we destroy God's creation it's similar to tearing a page out of the Bible. Now somebody who grew up as an evangelical Christian, you know, you're—one of the hallmarks of evangelicalism is a high regard for scriptures, a high regard for the Bible as sort of the supreme guide for living life, if you will. So I would never ever as a devout follower of Jesus deface or destroy a copy of God's word, and yet I realized at that moment that that's exactly what I was doing through the way I lived my life.[7]

The epiphany encouraged Merritt to reread the Bible, looking for guidance about how to steward the earth, and to read books on the topic of creation care. While searching online, he came across *Creation Care,* the magazine of the EEN. Realizing that its editorial offices were based nearby, he reached out to its staff and ended up meeting with Rusty Pritchard, who was then the magazine's editor, and Jim Jewell, one of its advisory editors.[8] During the meeting, Pritchard and Jewell handed Jonathan a packet that included a copy of the ECI's statement "Climate Change: An Evangelical Call to Action" (hereafter "Call to Action").[9] This turned out to be pivotal, for as Jonathan told me, "I went home and I thought about this whole ECI thing. And I said, you know, we could do something like that in Southern Baptist circles."[10]

As the son of a former president of the SBC, James Merritt, the younger Merritt was well positioned to gain support for such an effort from respected Southern Baptist leaders. Although he originally thought of it as just a "really neat blog post," he decided to take the project seriously, and enlisted the support of five "PhD-level folks" to draft the Declaration. Together, they created a document that they felt was "really sound, theologically, and that also was distinctly Southern Baptist in its tone and its approach."[11] Merritt then gathered signatures, reaching high-profile leaders—including Frank Page, president of the SBC at that time; Jack Graham, a former SBC president; and Danny Akin, president of SEBTS—as well as pastors and lay members from around the country. In the end he obtained over 750 signatures.[12]

In partnership with the EEN, Merritt initially planned to announce the Declaration in an event at the National Press Club. But the SBC's powerful political arm, the Ethics and Religious Liberty Commission (ERLC), was not pleased. Headed by Richard Land, the ERLC had backed the official skeptical resolution passed in 2007 and did not appreciate Merritt's attempt to publicize dissenting views. According to Merritt, when the ERLC learned of his project, its leaders pressured him to hand the project over. When he declined, they began a campaign of character assassination against him. "Suddenly [it] was 'Jonathan and his dark political ties' and 'He is in cahoots with this liberal group,'" Merritt recalled. A week before the planned release of the Declaration, a former seminary professor of his called to say that if he did not turn the project over to the ERLC, its leaders would be "forced to unleash the full scale of their arsenal" against him. Merritt was stunned: "I couldn't control that. I wouldn't be able to manage that. . . . I just got off the phone and wept. I was so overwhelmed. I mean, I'm a seminary student. . . . It was very political and very overwhelming."[13]

Canceling the National Press Club event, Merritt nevertheless contacted a *New York Times* reporter who announced the release of the Declaration in an article published on March 10, 2008.[14] In addition to the article in the *New York Times,* independent stories appeared in the major national news outlets National Public Radio, *Time* magazine, CNN, and the *Christian Science Monitor.*[15] Influential papers like the *Los Angeles Times* and the *Washington Post,* both among the top ten newspapers in terms of circulation at the time, also printed versions of an Associated Press story, as did at least thirty-five local newspapers.[16] With significant media coverage, the Declaration soon entered the scholarly literature as an example of evangelicals' growing interest in tackling the issue of climate change.[17] However, while Merritt himself viewed climate change as a real and serious problem, not all the signatories did.[18]

INTERPRETING THE DECLARATION

To begin to make sense of the existence of two opposing interpretations of the Declaration's position on climate change, it is helpful to turn to cultural studies. For cultural studies scholars, it is standard to assume that a text contains multiple potential readings. This is because readers have different assumptions, perspectives, and reading habits that shape how they interpret a text. Scholars have fruitfully applied this insight to a wide range of texts, but no one has yet applied it to texts produced by evangelical environmentalists, perhaps because activist texts are generally assumed to be created expressly for the purpose of advancing a particular position, and therefore to afford little opportunity for alternative readings. However, the unique conditions under which it was produced resulted in a more ambiguous text. Most significantly, its primary author was young, politically inexperienced, and under significant pressure from the powerful ERLC; he had also invited others whom he regarded as having greater theological expertise ("PhD-level folks") to help him craft the text. With their assistance, the Declaration made numerous appeals to traditional Southern Baptist values and commitments. While these appeals were intended to assure potential signatories that the text represented an authentically Southern Baptist voice, they also encouraged a reading of the text as consistent with the Southern Baptist tradition of independence and autonomy from opinions voiced by the secular world, including on the issue of climate change. Textual analysis is the ideal tool for understanding how these different readings were possible.

For textual analysts, a text's potential meanings become actualized whenever a reader encounters the text, yet this does not mean that each reader has a unique interpretation, creating infinite meanings. Rather, because certain groups have similar contextual knowledge, they tend to produce similar meanings from a particular text. As the cultural studies scholar Janice Radway has pointed out, readers thus interpret texts as members of "interpretive communities."[19] Here, I apply these insights to two interpretive communities, the first comprised of the national media and scholars, and the second comprised of Southern Baptist pastors who signed the Declaration and spoke publicly about this decision in two Southern Baptist–affiliated newspapers. That these two groups have different sociocultural positions is readily apparent; the national media and scholars largely belong to (and help produce) the secular world, whereas the pastors are immersed in both the culture of the Southern Baptist tradition and the broader culture of evangelicalism.

To disclose the different meanings that these two interpretive communities produced, I follow an analytical scheme developed by the cultural studies scholar Mikko Lehtonen, who recommends considering (1) how potential meanings are actualized in particular interpretive communities using different contextual cues and (2) the historical and cultural reasons that particular communities produce particular meanings.[20] Following this scheme for each interpretive community, I first detail the elements of the Climate Change Interpretation by analyzing the five independently reported news stories about the Declaration that were published in influential national media outlets (listed above), as well as the Associated Press story, which was widely republished. I then examine the contextual and textual factors that enabled this interpretation. Second, to explore the Stewardship Interpretation, I analyze the statements that Southern Baptist signatories made about the Declaration in articles published in *Baptist Press* (the SBC's official news service) and the *Christian Index* (the Georgia Baptist Convention's news service). I then examine the contextual and textual factors supporting this interpretation. Critically, I show that the different contextual clues each interpretive community employed directed their respective readers' attention to different aspects of the text, enabling the two contrasting interpretations to emerge.

To further illuminate how Southern Baptists interpreted the Declaration, I draw on the semi-structured interviews I conducted with signatories, the majority of whom were from Georgia. With about 18 percent of the state's population claiming SBC membership, Georgia is one of the denomination's strongholds.[21] Having been raised outside of Atlanta, the state capital, Merritt also had strong networks there, as indicated by the extensive coverage of the Declaration in the *Christian Index* and by a series of opinion pieces about it (not analyzed here) in the *Atlanta Journal-Constitution,* the city's flagship newspaper. Georgia pastors were therefore well positioned to be knowledgeable about the Declaration and the controversy surrounding it. That said, the pastors I interviewed are not a representative sample of the signatories, but rather a means of providing greater insight into the mindset of skeptical signatories.

As a final note, readers are advised to read the Declaration (contained in appendix B) before proceeding.

THE CLIMATE CHANGE INTERPRETATION

The Climate Change Interpretation entailed four basic assumptions: the Declaration was primarily about climate change; the Declaration favored action to halt climate change; the appearance of the Declaration indicated a shift in opinion among Southern Baptists on the issue; and this shift was an extension of previous efforts to green the evangelical tradition. Although this interpretation also appeared in the academic literature, it was largely produced by the national media, so I focus on that coverage here.

That the national news media assumed the Declaration was primarily about climate change is evident from their headlines, which asserted, for example, that "Southern Baptists Back a Shift on Climate Change" or that "Southern Baptist Leaders Take Unusual Step of Urging Fight against Climate Change."[22] Only the *Time* headline bucked this trend, linking the Declaration instead to the greening of evangelicalism.[23] Even though the Declaration also addressed environmental stewardship, most of the national news stories discussed the Declaration without reference to this aspect.

Most of the news stories also claimed or implied in their leads that the Declaration said climate change should be stopped. CNN's lead claimed, for example, that "several prominent leaders in the Southern Baptist Convention said Monday that Baptists have a moral responsibility to combat climate change."[24] National Public Radio, *Time,* and the *Christian Science Monitor* stopped short of writing that the Declaration said climate change should be stopped, but they implied that this was its message by describing it as a challenge to the SBC's official 2007 resolution advocating climate skepticism.[25] Five of the six stories supported their assertion that the Declaration represented a shift in opinion by referring to the line in the preamble stating that "our current denominational engagement with these issues have [sic] often been too timid."[26]

Finally, all the stories (except the very brief National Public Radio story) placed the Declaration within the larger context of the greening of evangelicalism. According to the *New York Times,* for example, "the Southern Baptist signatories join a growing community of evangelicals pushing for more action" on climate change.[27] Similarly, according to *Time,* "the position of Evangelicals in general (of whom Southern Baptists represent a sizable piece) has been swinging ever greener."[28] Only the *Christian Science Monitor* noted that there was any difference between the Declaration and previous examples of greening. Laudably, it quoted an author of the ECI's statement, David Gushee, as saying that the Declaration "doesn't go far enough."[29]

Contextual Factors Enabling the Climate Change Interpretation

The national news media formed the Climate Change Interpretation, as readers do, by using contextual clues to inform their understanding of the text's meaning and intent.[30] Probably the most subtly powerful of these contextual clues was the Declaration's genre. Readers expect declarations to announce an opinion or a position—perhaps one that is controversial—and to stand by it. The Declaration did not overtly embrace climate skepticism (as the official SBC resolution had), so readers in the national news media were led to expect that it would be making a strong statement in favor of the opposite position.

A second key contextual factor, explicitly referenced in most of the news stories, was the broader phenomenon of the greening of evangelicalism. The year 2008 was a time of particularly high expectations on the part of scholars, activists, and journalists about the green evangelical movement's potential.[31] That a narrative arc of ascension was already in place encouraged the national news media to interpret the Declaration as extending the curve. In addition, although only indirectly referenced in most of the news stories, the ECI—the most recent high-profile example of evangelical greening—also likely affected how the media read the Declaration. The ECI was widely regarded as a significant contribution to global warming activism in the United States, and the superficial similarities between its Call to Action and the Declaration invited a parallel reading of the two texts. In fact, the initiative associated with the Declaration was called the Southern Baptist Environment and Climate Initiative, the initials of which—SBECI—hinted that it was a Southern Baptist version of the ECI. Finally, that the Declaration's spokesman, Jonathan Merritt, clearly considered climate change to be a real and serious problem surely influenced the national news media's interpretation of the Declaration.

Textual Support for the Climate Change Interpretation

The contextual cues discussed above elicited certain expectations about the Declaration's meaning. This, in turn, led readers in this interpretive community to focus on the elements of the Declaration that supported this reading, while ignoring those that did not (the section below on the Stewardship Interpretation discusses the elements that did not support this reading).

Regarding the assumption that the Declaration favors action to halt climate change, the Declaration's second headline statement is that "it is prudent to address global climate change." Even though the word *address* can mean simply "to discuss," it was interpreted here to mean "stop" or "mitigate." Hence, assumptions about the Declaration's intent determined the preferred reading. The Declaration also cites "general agreement among those engaged with [climate change] in the scientific community" in this section. Questioning whether there is scientific consensus about the reality of anthropogenic climate change has been a key talking point for climate skeptics, so this reference to "general agreement" would seem to clearly place signatories among those who accepted anthropogenic climate change.[32] The Declaration later reinforces this impression by describing the scientific evidence in support of climate change as "substantial." In another apparent show of solidarity with climate activists, the Declaration resolves "to engage this issue without any further lingering over the basic reality of the problem or our responsibility to address it."

Two statements in particular supported the interpretation that the Declaration represented a shift in opinion. First, the preamble notes that some signatories "had required considerable convincing" to become persuaded of the importance of environmental and climate change issues. Second, the preamble notes that the denomination had previously been "too timid" on these issues. Both lines suggested that the Declaration represented a departure from the SBC's official stance.

Finally, a handful of phrases in the Declaration supported the assumption that it endorses action: its statement that "humans must be proactive" when it comes to climate change, its reference to the need for "concrete action," the title of its final section ("It Is Time for Individuals, Churches, Communities and Governments to Act"), and the "pledge to act" in the conclusion.

Clearly, there is textual support for the Climate Change Interpretation. This interpretation is not inherent to the text, however, but emerged via the support of the contextual factors mentioned above. With other contextual clues, a different interpretation emerged.

THE STEWARDSHIP INTERPRETATION

According to the Stewardship Interpretation, the Declaration is primarily about the need to take care of the environment and is agnostic about the reality of climate change (permitting a range of beliefs, including skepticism). The full extent of this interpretation within Southern Baptist circles is unclear, but it is remarkable that all the prominent signatories who were quoted in the Baptist-affiliated presses, except Merritt, articulated it. This matters because it was their participation that made the Declaration newsworthy. That the majority of those I interviewed also held the Stewardship Interpretation suggests that it may have been common outside these circles as well, further calling into question what the Declaration represented.

That the signatories who were quoted in the Baptist-affiliated presses considered the Declaration to be primarily about environmental stewardship rather than climate change is clear from numerous statements they made. J. Gerald Harris, the *Christian Index*'s editor (who originally signed, but later removed his name), wrote that "some have unfairly lumped . . . Jonathan Merritt . . . and the signatories of his initiative in with Al Gore and his global warming groupies, but Jonathan's Declaration has less to do with global warming and more to do with environmental stewardship."[33] SEBTS president Daniel Akin and then SBC president Frank Page (both signatories) expressed the same sentiment, arguing that the Declaration did not contradict the 2007 resolution.[34] Akin's assessment is particularly noteworthy given that, according to Merritt, he had helped craft the document. Similarly, according to Merritt's father, global warming is "a minor point in Jonathan Merritt's document."[35] In an interview with *Baptist Press,* Union University president David Dockery (another co-drafter) "said he believes the scientific data on global warming is 'inconclusive' and that 'at best,' global warming is neither a primary or secondary moral issue but a 'tertiary issue.'"[36] In an article-length interview published in the *Christian Index,* even Merritt argued that the Declaration is "*not about global warming,* or a political agenda, it's about creation care and Christian stewardship, a theology that we have always embraced."[37] Even when they were quoted in the national news media, all the men except Merritt spoke about environmental stewardship, rather than climate change. National news reporters, however, failed to notice the omission.

Interestingly, a number of signatories complained that the impression that the Declaration is about climate change had come not from the document itself, but from the national news media. They were aware, in other words, that two different interpretations were circulating. According to Page, "this issue has been brought to a point where it is an internecine debate. This has been caused, in part, by secular media misconstruing the very basis of this issue."[38] James Merritt argued that "if you set your preconceived notions aside and read the document carefully it does not take the alarmist position."[39] The younger Merritt alluded to the same problem, stating, "I have been surprised at the negative reaction by people who have clearly never even read the document. Many are just responding to what others have written about it."[40] According to an interview with Akin that was published in the *Christian Index,* Akin also "wishe[d] more people would read it before voicing their disagreement."[41]

Given that signatories defended the Stewardship Interpretation *after* the national news media had advanced the Climate Change Interpretation, it is possible that they developed the Stewardship Interpretation only as an attempt to quiet the controversy. However, several pieces of information I uncovered suggest this was not the case. First, ten of the eleven pastors I interviewed also held the Stewardship Interpretation.

None of them reported receiving any negative feedback from coreligionists about their decision to sign on (something I specifically asked about), so their adoption of the Stewardship Interpretation is unlikely to indicate a loss of courage in the face of public pressure. This implies that at least some Southern Baptists originally read the Declaration as supporting the Stewardship Interpretation. Second, it is implausible that as late as 2008 the group of savvy, seasoned SBC leaders quoted above would have been unable to anticipate that signing a statement advocating action on climate change would be controversial; the SBC had passed a resolution on the issue just the previous year, and climate change had been routinely in the news, especially after the United Nations' Intergovernmental Panel on Climate Change published its fourth assessment report in 2007. It seems more plausible that (as they themselves argued) they did not realize the statement would be interpreted as a piece of climate change advocacy.

Two further pieces of evidence support the inference that some signatories originally read the Declaration as supporting the Stewardship Interpretation. The first comes from comments that the *Christian Index*'s editor, J. Gerald Harris, made in an editorial about the Declaration. There Harris mentioned that he had chaired the committee that presented the 2007 SBC resolution urging climate skepticism; he therefore would have been very familiar with the dimensions of the climate change controversy. Nevertheless, he also admitted that he was "one of the original signatories" of the Declaration.[42] This clearly indicates that he did not initially read the Declaration as conflicting with his skeptical views on climate change. Second, one of the pastors I interviewed had been a close associate of Merritt's while he was developing the Declaration and had even discussed its ideas with Merritt while he was composing it. Yet even with this firsthand information, Pastor Paul felt comfortable signing on as a climate skeptic. Given his familiarity with the document—he could still quote lines from it—it is unlikely that he simply did not read it carefully, and given his cogent explanation for why he signed (discussed below), it is unlikely that he changed his mind about climate change after signing. Rather, it seems most likely that he simply read it as being compatible with climate skepticism.

All these pieces of evidence suggest that some Southern Baptists genuinely read the Declaration as being consistent with climate skepticism. In other words, the Stewardship Interpretation was not created as a defensive maneuver but emerged organically out of the experiences and perspectives that some Southern Baptist readers brought to the text.

Contextual Factors Enabling the Stewardship Interpretation

As with the Climate Change Interpretation, contextual factors enabled Southern Baptists to develop a particular interpretation, one in which the Declaration is compatible with climate skepticism. Two contextual factors enabled this interpretation. First, while the national news media had made sense of the Declaration in the context of the greening of evangelicalism, many of the Southern Baptist pastors I interviewed acknowledged that they paid attention not to the document's stance on climate change but to its theology and the theology of other signatories. Their primary concern in signing on, that is, was not to ensure that the document made a sound case for climate skepticism (for they were unfamiliar with the flashpoints of this debate) but to ensure that they were in good (i.e., conservative) theological company. Once they felt confident in this regard, they read the Declaration through the prism of the tradition's sense of embattlement with secular culture, a reading that was congenial to climate skepticism.

A number of interviewees emphasized that what they paid attention to when trying to decide whether to sign on was the author's theological stance. Pastor John, for example, had learned of the Declaration through coverage in the *Christian Index*. When I asked if he had had any hesitations about signing on, he did not reference the Declaration's stance on climate change but instead replied, "I wanted to read through [Jonathan Merritt's] theological stance and understand where he was coming from. I read his statement of faith and what he believed and then I really didn't have a problem with it." When I asked him to elaborate, he continued:

> I wanted to make sure that his statement—his theological statement—lined up with what I believe the Bible teaches on creation and evolution—on any number of topics—about the person of Christ, about any number of things like that. I wanted to make sure that I was comfortable with—and plus, he had pastors that I respect that had signed the pledge as well. So when I saw that, I was a little more comfortable with it.

For John, it was more important to know whether the authors and other signatories held conservative theological views than it was to ascertain the Declaration's specific views on the environment and climate change.

Similarly, when I asked Pastor Aaron if he recalled what went through his head as he was trying to decide whether to sign on, he replied without hesitation, "Did it stand for the faith that I believe in and did it reflect it clearly and concisely." Pastor Jim reported signing because being good stewards of the earth "makes good theological sense." When I asked him if he had taken any flak for signing (thinking that the flak would be related to announcing his support for taking action on climate change), he instead responded, "Oh sure. . . . There seems to be a group of people who believe that if you are concerned about the environment that somehow you must be very liberal in your theological position." Echoing my focus group findings about the importance of social factors, for the pastors I interviewed, the primary concern when deciding whether to sign the Declaration was whether it would harm their reputation as theological conservatives.

With the Declaration's conservative theological credentials established, pastors next focused on a line in the preamble that the national news media had almost completely ignored. After admitting that the denomination's engagement with the environment and climate change had "often been too timid," and that such a response might be seen as "uncaring, reckless and ill-informed," the Declaration states that "to abandon these issues to the secular world is to shirk from our responsibility to be salt and light." The reference to salt and light is an allusion to Matthew 5:13–16, which Southern Baptists and other evangelicals interpret as admonishing them to engage actively with secular culture on questions of morality and faith. As the ERLC explains,

> Jesus expects His followers to apply biblical principles to those things in our culture that destroy lives and tear families apart, things such as poverty and hunger, perversion of God's design for human sexuality, devaluation of human life at all stages, all manner of greed, and the lack of justice for all.[43]

Hence, while the sentence from the Declaration clearly encourages engagement, it does so in language that calls to mind Southern Baptists' long-standing battle with secular culture over issues like gay marriage and abortion. Given this background, it is less surprising that the Declaration itself mentions abortion no less than three times, even describing it and biblical definitions of marriage as "the most pressing moral issues of our day"—more pressing than climate change, in a document ostensibly *about* climate change! For those who viewed themselves as participants in such battles, calling for *engagement* on the issue of climate change could be interpreted to mean calling for a combative, distinctly Southern Baptist alternative to existing secular responses. Those who held the Stewardship Interpretation were thus like the national news media in that they were attuned to the conventions of the declaration genre—which leads readers to expect a strong statement of opinion—but what they understood themselves to be declaring was quite different.

Among the pastors I interviewed, Pastor Paul expressed this view most articulately. Upon learning that Paul was a climate skeptic, I asked how he felt about the Declaration's statement that "it is prudent to address global climate change." He responded, "What I understand it to say there is [that] we need to be in the discussion. Again, it's a question of are we going to relinquish this entire issue to the more liberal side of the political arena and just basically avoid it altogether. I think that's unhealthy." Paul's statement put into context a comment he had made earlier about the notion that the denomination's previous engagement had been "too timid": "I agree with [Merritt]—that Christians ought to be the most concerned for the environment because we believe it is designed by a Creator. . . . I believe, however, that we have, as a church and as a convention, almost relinquished that issue to the liberal platform." Obviously, his interpretation of the reference to timidity differed greatly from the one advanced in the national news media. Rather than indicating that a stronger stance was needed on climate change, Paul's decision to sign had been intended to convey frustration that his denomination had not yet publicly marked out a Baptist position on environmental stewardship. For him, signing the Declaration was a way for Baptists to enter the debate on their own terms.

Pastor Matthew, an itinerant evangelist whose ministry focused on marriage counseling, explained his decision to sign on to the Declaration in similar terms:

> I'm a very traditional, biblically based Christian and I realize there are traditional ideas of climate change that are not biblical, but I was encouraged to know that there are different ways of viewing climate change that are not as liberal as what, I guess, traditional educators would try to present. So I felt more comfortable with this.

When I asked what he meant by a liberal view of climate change, he responded, "Just the traditional global warming idea. It really doesn't seem to have a great deal of support in factual data and it sure doesn't match up with the Christian worldview." When I inquired whether he believed that climate change was man-made, he answered by critiquing

> the whole idea that is promoted through educational institutions that it's all man-made and that by our failure to embrace a green climate agenda, that

we are endangering our planet. I believe God created the world in which we live and that this is not a surprise to Him. And that He is ultimately in control, not our educators.

What had made him sign on to the Declaration, then?

I realized that these things are of a concern to many individuals and I think we do need to thoughtfully come to some conclusion about how we, as Christians, deal with the issue of climate change. So I am in favor of considering these matters and doing what we can do.

In other words, signing on to the Declaration was a means of expressing his support for beginning a conversation about a "biblical" (i.e., skeptical) view of climate change—not for taking action to stop it.

While this view does not necessarily reflect all 750 signatories' interpretation of the Declaration, a number of clues suggest that it was influential for the prominent signatories discussed here. When Akin explained his rationale for signing the Declaration, for example, he stated that "those of us who affirm the Word of God should be at the forefront of this discussion. . . . It is unconscionable for us to turn that area over to pantheists (those who worship nature) and liberal environmentalists."[44] Both Jonathan Merritt and his father employed similar rhetoric. According to the younger Merritt, "If we remain true to God's Word, Christians must with equanimity redeem our cause and make it our own. To leave these issues to secular environmentalists is to abandon our God-given responsibility to care for His planet."[45] Similarly, for the elder Merritt, "Christians should have a place at the table when it comes to shaping public opinion; instead we have abdicated our role in this discussion."[46] Even though the younger Merritt was likely attempting to leverage the language of embattlement for the benefit of encouraging Southern Baptists to endorse action on climate change, by deploying it in the Declaration he (I believe unintentionally) welcomed climate change skeptics into the fold. Rather than challenging the SBC's official views on climate change, for these individuals the Declaration appeared to offer the opportunity to push back against encroaching secular society in a new arena: the debate over climate change.

Textual Support for the Stewardship Interpretation

As with the Climate Change Interpretation, contextual factors elicited certain expectations about the Declaration's meaning, which led readers to focus on elements of the text that supported this reading, while ignoring those that did not.

The best way to understand how it was possible for signatories to read the Declaration as compatible with climate skepticism is to compare it with the Call to Action, the parallel document produced under the auspices of the ECI (reproduced in appendix C). Unlike the Declaration, the Call to Action articulates its position on climate change unambiguously. As a result, comparing the two documents highlights how moderate and ambiguous the Declaration is regarding climate change, key factors that enabled skeptics to read it as compatible with their position. Importantly, the Declaration's moderation is no accident. Comparing the two texts reveals that numerous phrases and, in several cases, full sentences that appear in the Call to Action also appear in the Declaration. In total, 19 percent of the wording is identical, and an additional 3 percent expresses the same idea in slightly different words. This suggests that the Declaration was created by revising the text of the Call to Action. Hence, the Declaration's tone should be interpreted as a deliberate and conscious departure from the document it was modeled after. Again, the changes were probably not intended to welcome climate skeptics, but rather to welcome Southern Baptists by creating something that was, as Merritt had stated, "distinctly Southern Baptist in its tone and approach." However, it was this distinctive tone that seems to have primed some Southern Baptist readers to view it through the prism of the tradition's long-standing sense of embattlement with secular culture.

Comparing the Call to Action with the Declaration

The two most important ways that the Call to Action and the Declaration differ are in the extent to which they acknowledge the human contribution to climate change and in the emphasis they place on climate change itself. In both cases, the Declaration is more moderate than its model, avoiding opportunities to state that climate change is primarily caused by human activities and focusing on environmental stewardship rather than climate change. Both of these editorial decisions enabled readers with a certain background to interpret the Declaration as compatible with climate skepticism.

Regarding the emphasis on drivers of climate change, the Call to Action clearly places itself in league with those who believe climate change is real, caused by human activities, and potentially dire in its consequences. It uses the modifier *human-induced* four times, specifying in the preamble, for example, that it is talking about "human-induced climate change," which it describes as "a real problem," and underlining its take on climate change again with the title of its first section: "Human Induced Climate Change Is Real." It also emphasizes the dire consequences of climate change with such statements as "Millions of people could die in this century because of climate change, most of them our poorest global neighbors."

By contrast, the Declaration references "environmental and climate change issues" as "real problems," but it never specifies that it means anthropogenic climate change. Instead of referencing dire consequences, it states that "humans must be proactive and take responsibility for our contributions to climate change—*however great or small*."[47] A number of the pastors with whom I spoke felt that climate change was mostly due to natural variation, but that human activities might have contributed in some small way. Thus, while the Declaration argues that climate change is at least partially caused by human activities, its wording also leaves room for those who believe that human activities play a minimal role to feel comfortable signing.

By avoiding reference to climate change's dire consequences, the Declaration also does not overtly place its signatories among those most concerned about climate change. Indeed, instead of a discussion of consequences, readers of the Declaration are treated to a cautiously worded discussion of the debate over the reality of climate change. Here the Declaration bends over backwards to be charitable toward those who doubt, conceding that "this is an issue where Christians may find themselves in justified disagreement about both the problem and its solutions." Similarly, it later describes those scientists who deny anthropogenic climate change as "sincere and respected." In another line that would have resonated with skeptics, it states that "we recognize that we do not have any special revelation to guide us about *whether* global warming is occurring and, *if* it is occurring, *whether* people are causing it."[48] Grammatically, the sentence concerns the ability of the Declaration's nonscientific authors to comprehend climate science, but the use of the terms *whether* and *if* also gives the impression that the authors are not convinced it is occurring, for anyone inclined to read it that way. In contrast to the ECI's Call to Action, then, which leaves no doubt about its stance regarding climate change, the Declaration introduces numerous elements that could be read as compatible with—or at least tolerant of—climate skepticism.

A second important way in which the Declaration is far more moderate than the Call to Action has to do with its focus. While the Call to

Action focuses exclusively on climate change, the Declaration chooses to emphasize environmental stewardship, leaving climate change to be discussed substantively only in one section (a section that, as described above, seems tolerant of skepticism). Pursuant to this emphasis, it is the environment, not climate change, that is listed first in the Declaration's title and that is the main subject of the Declaration's first, third, and fourth sections. "Climate issues" are briefly referenced in the third section but are not differentiated from environmental issues. The last section, which is arguably the most rhetorically powerful in that it discusses the need to act, does not mention climate change at all, but instead refers generically to environmental issues, care for the earth, and biblical stewardship. In contrast to the Call to Action, then, the Declaration is not clearly focused on climate change, leaving room for alternative interpretations of its intent.

Importantly, what the Declaration says about environmental stewardship was neither new nor controversial at the time of its release. Its strongest statement is that "there is undeniable evidence that the earth . . . can be damaged by human activity"—hardly a revolutionary assertion in the year 2008. Even this is softened by the admission that human activity can sometimes be "productive and caring." Historical context makes the Declaration seem even more moderate, for official SBC resolutions approved in 1970, 1983, and 1990 had addressed the same topic in stronger words and without such caveats. Arguably the most mildly worded of these, the 1990 resolution "On Environmental Stewardship," still comes off stronger than the Declaration by criticizing "the destruction of the created order" and "human extravagance and wastefulness . . . and general misuse of creation."[49] Thus, the Declaration not only addresses a topic that had already been approved through official channels, but it does so in even milder language than its predecessors. Except for the climate change piece, it could arguably have been viewed as indicating that Southern Baptists were backing away slightly from their historical commitments, rather than pushing for change. This, too, would have seemed congenial to those who viewed the environmental movement—not to mention climate change—with suspicion.

Reading the Declaration through the Lens of Embattlement

That the Declaration is ambiguously worded with regard to climate change and seems to be primarily about environmental stewardship enabled Southern Baptists who were climate skeptics to find ways to read the rest of the document as congenial to their position. Here their familiarity with Southern Baptist history and culture was key, for it disclosed readings that were invisible to outsiders.

First, as noted above, the Declaration uses the subheading "It is prudent to address global climate change." Yet the term *address* does not actually indicate that action is preferred, a distinction that seems to have enabled some of the skeptical pastors I interviewed to interpret the Declaration as saying that Southern Baptists need to "address" climate change by adding their (skeptical) voices to the national conversation about it.

A second example comes from the Declaration's line about the denomination being "too timid" on climate and environmental issues. In the national news stories, this was interpreted as evidence that Southern Baptists were shifting away from their skeptical position on climate change. Yet a brief consideration of Southern Baptists' historical engagement with environmental issues yields another plausible interpretation. In addition to the official 2007 resolution urging caution about climate change, the SBC had resolved in 2006 to "resist alliances with extreme environmental groups whose positions contradict biblical principles . . . and [to] oppose solutions based on questionable science."[50] Thus, the assertion in the Declaration that the denomination had been too timid in engaging with environmental and climate change issues could conceivably be interpreted to mean that it had not gone far enough in opposing green radicalism—or simply that it had not inserted its (skeptical) voice aggressively enough into the national conversation. This is quite different from the national news media's interpretation of this statement as saying that the denomination had not embraced environmental concerns enthusiastically enough.

A third example is another much-quoted line from the Declaration's final section: "We pledge . . . to give serious consideration to responsible policies that acceptably address the conditions set forth in this declaration." Although the national news media interpreted this as a commitment to take action to halt climate change, it was also possible to read it as compatible with climate skepticism. First, the sentence never states that the "conditions" include anthropogenic climate change, instead referring readers back to ambiguous statements made earlier. Second, unlike the Call to Action, the Declaration never suggests what would constitute "responsible" policies, so the intended meaning of that word is left open to the reader's interpretation. For climate change skeptics, this could be taken as a plea not to go too far by adopting "radical" policies. The pastors I interviewed repeatedly stressed their preference for the "commonsense" and the "practical" solutions (often those that did not require government intervention)—options I am sure they viewed as responsible. Similarly, the 2007 SBC resolution argues that proposals to regulate greenhouse emissions are themselves "very dangerous"—the opposite of responsible. Thus, Southern Baptists reading the document would not necessarily have interpreted the Declaration's support for "responsible" policies as support for any of the policy solutions that were then on the table. Finally, the word *acceptably* diminishes the sentence's force significantly, suggesting that the conditions should be addressed only in a way that the signatories agree with—yet another check that the signatories placed on policymaking. For those so inclined, it could even be read as an oblique criticism of policy options that were currently on the table, suggesting that they were *un*acceptable.

To sum up, contrary to my expectation going in, my research ended up underlining the embattled mentality's significance as a barrier to activism on anthropogenic climate change. But if this is in fact the case, it raises at least one important question. Why would prominent Southern Baptist leaders and others feel called to publicly endorse a declaration about something as anodyne as environmental stewardship? Understanding why requires considering the broader picture of evangelical engagement with climate change. As I will argue in Part Two, the "greening of evangelicalism" narrative that developed in the national news media between approximately 2005 and 2008 helped create a situation in which evangelical opponents of action felt they could not simply let their position be silence or uncoordinated, off-the-cuff dismissals. Rather, they had to fight back—that is, to actively promote skepticism as the Christian position on climate change—and their strategy for doing so entailed redefining environmental stewardship so as to exclude concern about anthropogenic climate change. It is this campaign, I argue, that ultimately helps explain the major finding of Part One, that climate skepticism had become embedded in white traditionalist evangelicals' sense of identity.

• • •

PART ONE: SUMMARY AND IMPLICATIONS

Environmentalists have suspected for decades that evangelicals are unconcerned about the environment because of their end-time beliefs.

What I learned from speaking to evangelicals in Georgia was that this suspicion does point to something real: my informants were generally unconcerned about the environment. Unpacking that lack of concern revealed an interesting twist: they generally agreed with environmentalists about the importance of maintaining a healthy environment and about the enjoyment that could be found in nature, but they felt that solving environmental problems should not require any special effort but could and should be accomplished simply by individuals employing common sense.

As for end-time beliefs, a small number of those I met (the hot millennialists) fit the classic end-time apathy hypothesis, being unconcerned (in fact, excited) about signs of planetary decline because of their belief in Jesus's imminent return. However, these individuals tended to be politically disengaged. The majority of my informants (cool millennialists) were unconcerned for a different reason, namely because they viewed climate change through the prism of their sense of embattlement with secular culture. These informants' attitudes about climate change were shaped not only by their religious outlook, but by what they had heard from secular climate skeptics, whose rationales for skepticism often resonated with their own suspicions about secular society. In addition, to the extent that cool millennialists' climate skepticism was tied to their faith, it was sustained by their faith *community*—via norms of attention and moral relevance and through identity work—which itself found meaning in the quest to return Christianity to its rightful place in American society. And as we have just seen, a number of Southern Baptist leaders shared this same view when it came to interpreting climate change.

As for implications, if climate change skepticism has indeed become tied to the embattled mentality for a certain group of evangelicals, the traditionalists, then this explains why evangelical climate activists have had so much difficulty making headway among the laity. Sociologically speaking, the sense of embattlement with secular culture is a key reason evangelicals have been able to build such a strong, culturally influential movement, for as Christian Smith and his colleagues have compellingly argued, it is evangelicals' sense of distinction from and tension with the non-evangelical world that has fostered the movement's cohesion, increased its resource mobilization, and enhanced its ability to retain members, ultimately enabling it to thrive in a modern, pluralistic setting.[51] That climate change attitudes have apparently become tied to the very beliefs that have made evangelicalism resilient means that environmentalists hoping evangelicals would join forces with them faced stiff cultural head-winds indeed.

An Embattled Perspective on Climate Change: Part of a National-Level Trend?

Two further questions emerge from these findings. The first is one that came up whenever I discussed my findings with others: Did the embat-tled mentality play a role only in Georgia or was it part of a broader, national-level trend? The question is worth asking not only for aca-demic reasons. Because of the evangelical movement's cultural and political influence in the United States, evangelical opposition to action on climate change could prevent or at least significantly impede the pas-sage of climate legislation in the United States. In turn, because the United States is generally viewed as a key participant in global efforts to curb greenhouse gas emissions, its decision not to participate would hinder the global effort. Hence, American evangelicals were relevant players in the global effort to develop a response to climate change.

Not having conducted a nationally representative study, I cannot definitively answer whether traditionalist evangelicals around the coun-try see climate change in similar terms as did my informants in Georgia, but I do see several lines of evidence that point strongly to this conclu-sion. First, it is clear from Christian Smith and his colleagues' work, as well as from other studies, that the embattled mentality is a national phenomenon.[52] This at least satisfies the precondition for evangelicals around the country to be able to view climate change through this lens. Second, every qualitative study I have reviewed—conducted in different parts of the country, including the South, Southeast, Southwest, and West—describes comments that were virtually identical to what I heard in Georgia. In focus groups conducted in Georgia, South Carolina, and Tennessee, for example, Katharine Wilkinson

> repeatedly heard a core justification to inaction: People do not have the power to cause or to rectify climate change. . . . [For many of my partici-pants] affecting God's creation in such a profound way is simply beyond the bounds of human capacity, and belief in climate change implies a misplaced confidence in the power of human beings.[53]

In a study conducted with churchgoers belonging to a number of dif-ferent evangelical denominations in Dallas, Texas, Wylie Carr and his colleagues noted that a third of their interviewees referenced the same idea without prompting—that God is in control of the climate. Moreover, the informants who stated that their faith affected their opinions about climate change (a majority in his sample) all shared the view that "environmentalists and climate change advocates misunderstand the created order and hence the proper relationship between God, humans and nature."[54] This summary statement reflects embattled evangelicals' framing of environmentalists as cultural opponents. In another reference to the embattled mentality that matches what I encountered in Georgia, a majority of the respondents in Carr's study dismissed the possibility of anthropogenic climate change by arguing that God, rather than humanity, is in control of the end times.[55]

In their qualitative study of black and white evangelicals from two churches in the Southwest, Jared Peifer and his colleagues found that "evangelical belief in God's sovereignty was frequently raised," describing one white evangelical informant as saying that "[God's] got this global warming thing under control, so I don't really need to—it doesn't make me lose sleep, because God's got it."[56] Like my informants, the participants in their study criticized environmentalists and scientists "for espousing a secular version of End Times" when talking about climate change.[57]

Finally, in research on pastors and laypeople in Arizona and California who attended the conservative evangelical Calvary Chapel association of churches, Bernard Zaleha found that the idea that God was in control "permeates sermons and conversations among members."[58] Like mine, his informants worried about secular attempts to take God out of the equation, criticized anyone who thought humans, rather than God, could affect environmental outcomes, and viewed environmentalists as spiritually threatening. All four views reflect the sense of embattlement with secular culture.[59]

Survey research also suggests that the sense of embattlement shapes evangelicals' attitudes toward climate change around the country. In a nationally representative survey conducted in 2016, the Yale Program on Climate Change Communication found that 30 percent of self-identified evangelical and born-again Christians affirmed a view that captures well the connection between the embattled mentality and climate change: "God controls the climate, therefore people can't be causing global warming."[60] Interestingly, an even higher percentage of Tea Party members, a group that overlaps with supporters of the Religious Right, affirmed the same view: 38 percent of Tea Partiers agreed with this view, while according to Pew Research, 69 percent of those who agreed with the conservative Christian movement also agreed with the Tea Party.[61]

The biannual National Surveys on Energy and the Environment, which include questions about climate change, offer further evidence of a national-level trend. When evangelical respondents (measured by self-identification) who did not believe the earth's climate was changing were asked to give the *primary* reason for their skepticism, 22–31 percent cited their religious beliefs in 2012 and 2013.[62] In addition, a review of the comments that respondents made during the years 2010–15 directly implicates the embattled mentality, underlining that, as other studies have found, a significant minority of evangelicals in that period viewed climate change in religious terms. The majority referenced the same narrative I had encountered during my fieldwork, stating, for example, that they were skeptical that the climate was changing because "God handles the weather" or "it is all in God's hands." Notably, only one evangelical respondent between 2010 and 2015 described climate change as a sign of the end times.[63]

It is important to acknowledge that while the "God is in control" rationale was a major theme in the qualitative studies and survey research just described, in none of them was it the only rationale informants offered for their skepticism. Certainly, as my own work has shown, many factors are responsible for evangelicals' skepticism toward climate change. But, taken together, these studies do appear to suggest that the subculture of embattlement shapes the climate change attitudes of up to a third of American evangelicals.

This brings me to the second question that the findings presented thus far raise: Did evangelicals around the country just happen to spontaneously apply their Christian worldview to the issue of climate change in the same way? Or had there been more specific forces operating to encourage this particular interpretation? On one hand, the similar ways in which evangelicals around the country were speaking about climate change implied some kind of coordinated effort to shape their opinions. Moreover, the historical record clearly shows that evangelicals arrived at their position on other issues, such as abortion, via the efforts of leaders.[64] On the other hand, the terms evangelicals around the country were using ("God is in control") were generic enough that it was possible to imagine that people had arrived at similar interpretations independently. Moreover, in a movement as large and decentralized as evangelicalism, how would it be possible to popularize one specific message about climate change, and to do so with no one in the secular world noticing?

This brings us to Part Two, which is devoted to explaining *how* traditionalist American evangelicals converged on a particular interpretation of climate change as a front in the war against secularism. I argue that it is not a coincidence that evangelicals around the country hold these views; rather, a major reason they hold them is because leaders and pundits associated with the Christian Right mounted a powerful media campaign encouraging them to do so.

PART TWO

How Skepticism Became the Biblical Position on Climate Change

CHAPTER 7

Preaching the Gospel of Climate Skepticism

On November 30, 2015, readers of the *New York Times* would have encountered a front-page picture of crowds gathered in Lisbon on the eve of the United Nations Climate Change Conference in Paris, France. Looking sober, participants held a sign reading simply, "There's no Planet B." The *Times* story noted that "recent scientific reports have concluded that the first effects of human-caused climate change have already started to sweep across the earth, from rising sea levels . . . to savage heat waves." It also noted that 2015 was on track to be the hottest global year on record—topping the record set the previous year.[1]

The *Times* story presented climate change as real, serious, and potentially catastrophic if not addressed. But for the millions of American evangelicals who regularly seek a Christian perspective on the news, the story they would have encountered that day was much different. For example, in a radio feature that claims eight million weekly listeners, John Stonestreet, president of the Colson Center for Christian Worldview, criticized U.S. participation in the accord as naive and asserted that climate models had been proven to be inaccurate. But even worse than the agreement's technical deficiencies, according to Stonestreet, was the worldview it represented. President Obama had described the accord as showing "that the world has both the will and the ability to take on this challenge." Stonestreet critiqued this statement as demonstrating a commitment to a "sort of secular technocratic salvation history" in which "there is no problem beyond the reach of human ingenuity." He admitted the importance of "proper stewardship of creation" but emphasized that it was improper to "[insist] that the steward—us—has salvation-like powers to restore the creation. . . . For that, only a real salvation history with a real savior will do."[2] In Stonestreet's telling, it was not an international agreement that was needed, but Christianity.

Those who preferred to get their news via television rather than radio might have caught a story on the Christian Broadcasting Network (CBN) about the conference. Founded in 1960 by the charismatic preacher Pat Robertson, CBN had grown from airing on a small local television station to allegedly reaching one million daily viewers nationwide via its flagship program *The 700 Club,* hosted by Robertson.[3] The story that aired on *CBN News* that day presented a different take than the one heard on Stonestreet's radio feature *BreakPoint* (see also below) but was equally skeptical. The CBN reporter described Obama's claim of a general scientific consensus on climate change as something "that's been shown to be wrong in peer reviewed journals," citing a *Wall Street Journal* opinion piece by two men affiliated with the skeptical Heartland Institute.[4] The only supporter of the agreement the reporter quoted—besides Obama and, briefly, the United Nations secretary general—was Al Gore. Rather than offering Gore's take on why an international agreement was needed, however, the reporter chose to highlight Gore's impolitic admission (or so it appeared in this context) that the goal of the negotiations was to create an agreement that could be enacted without U.S. congressional approval. The reporter went on to present Senator James Inhofe's and other Republican politicians' negative assessments of the Paris conference, concluding dismissively that "world leaders are spending a lot of money . . . to hold a conference on an issue that fewer and fewer Europeans and Americans care about."[5] Viewers would be left with little doubt as to whether an international climate agreement was worthwhile or necessary.

BreakPoint and *CBN News* represent only a small slice of what is known as the evangelical mass media, a "vast media infrastructure" catering to evangelical Christians that includes television, radio, and digital media.[6] Much of the content presented in the evangelical mass media is devoted to communicating nonpolitical messages.[7] But I found that among the programs that regularly consider political topics—news programs that report on current events "from a Christian perspective" and what Dunn and Tyler call "advocacy talk shows," or programs that present an evangelical perspective on the news in a discussion-based format—the coverage of climate change from 2006 to 2015 was heavily tilted toward skepticism.[8]

I argue that the skeptical communications included in these evangelical news and talk programs, together with similar material that appeared in print and digital media catering to conservative evangelicals, help explain why evangelicals from across the country converged on a similar interpretation of anthropogenic climate change as incompatible with their faith. To be clear, these communications do not appear to have entirely created the phenomenon of evangelical climate change skepticism, for most polls conducted in the early 2000s show that white evangelicals were already less concerned about climate change than other Americans.[9] What they did was subtler, but in the end still powerful. As the communications scholar Mark Ward Sr. has observed, the evangelical mass media are a critical means by which modern evangelicals—distributed throughout the country in tens of thousands of churches and denominations—"ritually create community."[10] Adding climate skepticism to this communal conversation helped hitch it to the powerful engine of evangelical subcultural identity. The mocking and dismissive coverage of climate advocates and advocacy that pervaded these programs powerfully conveyed the message that climate change was an issue that belonged to evangelicals' enemies, the secular elites.

THE "BIG FOUR" AND THE CAMPAIGN TO DISSEMINATE CLIMATE SKEPTICISM

Since one conclusion following my research on the SBECI was that the embattled mentality was likely linked to climate skepticism within the larger evangelical world, I decided to see if this discourse could be discerned at the elite level as well. I started with supporters of the ECI, examining their books and public statements to see if they had distanced themselves from the language of embattlement. I found that while promoting the ECI a number of them had in fact explicitly called for a less combative tone. Intrigued, I decided to see how individuals who had publicly opposed the ECI described their stance. While looking for statements they had made about climate change, I discovered something else: that many of these signatories had been extremely active in promoting climate skepticism to the evangelical laity.

Although evidence of their efforts was easy to come by, constructing a comprehensive overview of those efforts was difficult. The leaders and pundits who had promoted climate skepticism used a dizzying array of media and media platforms, and different individuals contributed in very different ways. Tim LaHaye cowrote a series of novels in which climate change advocates were portrayed as allies of the Antichrist, for example, while Chuck Colson primarily used his radio program and syndicated column to disseminate skepticism.[11] My solution was to focus on the efforts of the *Big Four,* a term coined by former NAE president Ted Haggard to refer to evangelical leaders who head or headed major media ministries.[12] Since these men had access to large audiences, examining their work captures the basic building blocks of what I have come to think of as a campaign.

By calling these activities a campaign, I mean that allied actors disseminated similar messages across multiple media platforms in a parallel, and at least loosely coordinated, fashion. Underscoring the point about allied actors, though I cannot discuss their efforts in detail here, politically conservative Catholics, most notably Phyllis Schlafly and Paul Weyrich, also contributed to the climate skepticism campaign.[13] These efforts appear to have been neither completely coordinated in the sense that a leader assigned specific roles to certain individuals, nor completely decentralized with each actor proceeding autonomously. However, as table 2 shows, most of the individuals who helped promote climate skepticism to evangelical audiences—with the important exceptions of Robertson, Falwell, and Ham—have been publicly associated with an organization known as the Cornwall Alliance (whose role in the campaign is discussed in chapter 8).

As for who the Big Four are (or were, in the case of those deceased), most Americans will recognize them: Pat Robertson, Chuck Colson, James Dobson, and Jerry Falwell. In addition to the Big Four, I looked at D. James Kennedy, who similarly merged a successful media ministry with conservative political activism, and his collaborator (on the issue of climate change at least) Ken Ham. Because Falwell, Kennedy, and Colson all passed away in the midst of the campaign (Falwell and Kennedy in 2007, Colson in 2012), I have also tracked efforts initiated by their successors. When possible I draw a still wider circle, showing how a leader's efforts were connected to other projects headed by collaborators or associates.

The period I investigated was 2006 (when the ECI launched) through 2015 (the most recent full year for which records were available when I began the analysis). As for the campaign's origins (so far as I could determine via the digital archives that were available), prior to 2006 very little was said about climate change on the programs I reviewed, and shows like Colson's and Ham's did not devote significant coverage until 2008.[14] In the next chapter I will explain this curious timing.

My account is not exhaustive, but it reveals what I believe to be the major elements of the campaign.[15] Tables 2 and 3 provide an overview of my analysis. Although appendix A provides further details about my methods, two comments about the tables are in order here. First, while I categorized stories according to whether they were skeptical, accepting, neutral, or unclear with regard to their stance on climate change, as table 2 indicates, only Robertson's programs fell into multiple categories. Hence, for the sake of simplicity, I included a column in the table only for the percentage of stories that were skeptical, omitting the other categories; the breakdown for the non-skeptical stories on Robertson's *CBN News* and *The 700 Club* is given below. Second, I was not able to review every program over the entire period, because some were not archived online. However, the consistency with which the programs I did investigate discussed climate change from a skeptical perspective suggests that accessing the data from those missing programs would not alter the picture presented here.

In the picture that emerges, Robertson's CBN, Christian talk radio, and online news play particularly important roles in spreading the message far and wide. Furthermore, the messages disseminated in this campaign match what I and others have encountered among lay evangelicals, suggesting how climate skepticism may have become linked to the embattled mentality—and, more generally, to faith—for a subset of theologically conservative Christians.

As for the campaign's effects, the key takeaway points are as follows. First, by addressing climate change in media that cater specifically to evangelicals, and by presenting it alongside issues like abortion, prayer in schools, and creationism, participants in the campaign made it clear that climate change was an issue that evangelicals should care about. In other words, they *transformed climate change into a religious issue,* symbolically on par with abortion or religious liberty. Second, while their communications tended to present secular rationales for climate skepticism, they also often *explicitly presented skepticism as the biblical view on climate change,* making it an issue, like abortion, where (they implied) the Christian stance was clear.

Finally, the campaign was *comprehensive.* Although I highlight the dissemination of skepticism via mass media, it will also become clear in what follows that participants utilized the machinery of conservative evangelical political engagement to reduce evangelical support for

TABLE 2 KEY CHRISTIAN RIGHT LEADERS' AND PUNDITS' USE OF TV AND RADIO TO DISSEMINATE CLIMATE SKEPTICISM, 2006–2015

Leader/pundit	Program	Audience	Number of stories or sermons[a]	Percent skeptical[b]	Period examined	Cornwall Alliance affiliation?
		Television				
Pat Robertson	CBN News, The 700 Club	1 million (daily)	58	79%	2006–2015	No
Jerry Falwell[c]	The Old-Time Gospel Hour	12 million (potential)	≥3	100%	2006–2007	No
		Radio				
Chuck Colson	BreakPoint	1,200 stations	22	100%	2009–2010	Yes
Ken Ham	Answers with Ken Ham	900 stations	28	100%	2008–2015	No
Janet Parshall	In the Market with Janet Parshall	372 stations	23	100%	2010–2015	Yes
Tony Perkins	Washington Watch/Live with Tony Perkins	300 stations	19	100%	2014–2015	Yes
David Wheaton	The Christian Worldview	225 stations	6	[100%?][d]	2008–2015	No
David Barton	WallBuilders Live!	200 stations	12	100%	2008–2015	Yes
Kevin Boling	Knowing the Truth	1 station	24	100%	2009–2015	Yes
Jerry Newcombe	Vocal Point	Online	6	100%	2014–2015	Yes
Kevin Swanson	Generations Radio	4 stations, online after 2011[e]	19	100%	2006–2015	Yes

Limited archives

James Dobson	Focus on the Family (radio)	2,000 stations	≥1	n/a	2006–2015	Yes
D. James Kennedy[f]	Truths That Transform (radio)	700 stations	≥1	n/a	2006	Yes
Jerry Falwell	The Old-Time Gospel Hour (radio)	Nationally syndicated[g]	≥3	100%	2006–2007	No

[a] Archives for which several weeks or more were inaccessible are denoted "≥" to indicate that further stories may have run during the missing weeks. Other archives are presumed to be complete.

[b] See appendix A.

[c] Falwell died in 2007, precluding further efforts.

[d] Archives were not available for the first five programs, but since E. Calvin Beisner was the guest on all of them it is reasonable to infer that the discussions would have presented climate change from a skeptical perspective. The last segment, which aired on May 15, 2015, was an interview with Ken Ham. It is available online, and I confirmed that Ham repeated his view that the climate is not changing as a result of human activities.

[e] By 2006, Generations Radio was airing on two stations in Colorado and two in Iowa. In January 2011, with online downloads outpacing AM/FM radio listenership, it transitioned to broadcasting completely online (Generations Radio, "Our Story").

[f] Kennedy sustained a heart attack in late 2006, likely limiting further efforts. The one program about climate change listed in the table is Beisner's interview with Kennedy, which later became the booklet Overheated. The book itself does not state when this interview occurred; my best guess is that it was in 2006, but after reviewing every entry available on Archive.org in 2006, I was not able to confirm this. I was able to determine that Beisner appeared on the program at least twice during this period, on June 20 and 21, with the topic "The Roots and Fruits of the Environmental Movement." It is unclear, however, whether these broadcasts were focused on global warming. Beisner did not recall whether these were the interviews that became the booklet Overheated (email communication, November 6, 2017).

[g] My efforts to determine how many stations Falwell's radio program aired on in the mid-2000s were unsuccessful. A dated figure from 1981 is 500 stations (Boodman, "How Falwell Raises His Millions").

TABLE 3 KEY CHRISTIAN RIGHT LEADERS' AND PUNDITS' USE OF ONLINE, PRINT, AND OTHER MEDIA TO DISSEMINATE CLIMATE SKEPTICISM[a]

Leader(s)/pundit(s)	Venue of skeptical communication(s)	Reach[b]
	Internet	
Chuck Colson	ChristianPost.com	1.1 million
	Crosswalk.com (Salem Web Network)	
Jerry Falwell	WND.com	1.3 million
Jerry Newcombe	ChristianPost.com	1.1 million
	WND.com	1.3 million
Pat Robertson	CBN.com	1.3 million
Ken Ham	AnswersinGenesis.org	740,000
	Magazines	
Ken Ham	*Answers*	70,000 subscribers[c]
Marvin Olasky and Joel Belz	*World*	100,000
	Books	
Wayne Grudem	*Politics according to the Bible* (2010)	
Ken Ham	*New Answers Book* (2010, 2013 editions)	
D. James Kennedy and E. Calvin Beisner	*Overheated* (2007)	
D. James Kennedy	*How Would Jesus Vote* (2008)	
Harry Jackson Jr. and Tony Perkins	*Personal Faith, Public Policy* (2008)	
James Wanliss	*Resisting the Green Dragon: Dominion, Not Death* (2011)	994 (low estimate)[d]

Other media

David Barton	*Science, the Bible and Global Warming* (CD, 2008)	
Many, including E. Calvin Beisner, David Barton, Bryan Fischer, Richard Land, Janet Parshall, and Tony Perkins	*Resisting the Green Dragon* (DVD, 2010)	931 (low estimate)
Ken Ham, D. James Kennedy, Richard Land	*Global Warming: A Scientific and Biblical Expose* (DVD, 2008)	

[a] This table summarizes the efforts described in this chapter. It is not an exhaustive list of the many media platforms via which climate skepticism has been communicated to evangelical audiences, nor does it represent the entirety of the effort by leaders and pundits associated with the Christian Right. Only websites with significant traffic are listed.

[b] The reach of websites is listed in terms of average monthly visitors from organic search (mobile and desktop), estimated via a SEMRush analysis conducted July 28, 2017. The mobile estimates are based on the number of users expected in the next month if average monthly organic traffic stays relatively the same ("organic traffic" refers to the traffic that comes from sources other than clicks via paid advertising). These figures should be interpreted as indicating the sizeable reach of these websites, but it is important to note that these figures do not include traffic from social media, referral sites, and other sources. I was unable to determine the reach for Crosswalk.com in 2017 via SEMRush and have chosen to leave the field blank rather than to insert an estimate calculated at a different time. Crosswalk.com itself claims 6 million monthly unique visitors (see p. 174). Magazines are listed in terms of circulation as reported by the publisher. Book sales are not listed because the publishers I contacted (except for the Cornwall Alliance) declined to share their figures.

[c] *Answers* claims 200,000 bimonthly readers but reports just 70,000 subscriptions (Answers in Genesis, "Answers Magazine," "2017 Media Kit").

[d] This estimate is based on an inventory conducted in 2015. The estimate reflects only online sales, excluding conference sales and other promotional events (Nancy Rogers, director of Donor Development for the Cornwall Alliance, email communication with author, November 16, 2015).

action on climate change. To be clear, not all evangelical media outlets participated in this campaign; some remained neutral or ignored the topic.[16] Nevertheless, those who did participate had a major impact by virtue of the thousands to millions of viewers and/or listeners they were able to reach. In media that cater to conservative or traditionalist evangelicals, insofar as I have been able to determine, skeptical voices were overwhelming.[17]

PAT ROBERTSON AND THE CHRISTIAN BROADCASTING NETWORK

In terms of audience size and efforts, two figures stand out as major contributors to the gospel of climate skepticism. Pat Robertson and Chuck Colson both have (or had, in Colson's case) the ability to reach millions of people, Robertson primarily via television and Colson primarily via radio. Each has added his unique imprint on the evangelical conversation about climate change. Robertson presents a particularly interesting case because, while his personal views about climate change are inconsistent, since about 2008 his network has leaned heavily toward skepticism in its news reporting and commentary.

Robertson was a key player in the rightward shift of evangelicals in the last quarter of the twentieth century, known for bringing together a diverse coalition of conservative Christians who viewed doctrinal issues as less important than attaining political influence. He also became a successful televangelist, attracting more than seven million viewers per week to *The 700 Club* by 1985. The show's audience had diminished to 828,000 by 2005, but that was largely due to the cable revolution, which fragmented audiences, rather than to declining interest in religious broadcasts.[18]

Compared to some of those discussed below, Robertson's environmental views are difficult to pin down. Robert F. Kennedy Jr., the attorney and environmental activist, has argued that Robertson played a key role in drumming up evangelical anti-environmentalism in the 1990s by making anti-environmentalism a "principal theme" on CBN's talk shows, news hours, and documentaries.[19] But Robertson has also devoted some positive coverage to environmental issues on *The 700 Club,* including a friendly interview with Al Gore in 1992 when Gore was promoting *Earth in the Balance,* a book whose main focus was the potentially dire consequences of climate change.[20] In the interview Robertson appeared genuinely concerned about environmental issues, including climate change, and even concluded by endorsing Gore's book as something that would make "a tremendous contribution."

When the controversy over evangelical engagement with climate change erupted in 2006, Robertson seemed ambivalent. In October 2005, while hosting a skeptical Senator Inhofe on *The 700 Club*, Robertson had added his own warning against worshipping the earth.[21] By August 2006, however, with temperatures reaching record highs, Robertson seemed to have changed his mind, announcing on *The 700 Club* that proponents of action on global warming were "making a convert out of me. . . . We really need to address the burning of fossil fuels." Despite this comment, Robertson's spokeswoman later claimed that he was undecided.[22] In March 2008, he seemed to double down on acceptance by appearing in a TV ad sponsored by Al Gore's Alliance for Climate Protection.[23] Sitting next to Al Sharpton, Robertson said that he and Sharpton disagreed about everything except "our planet. Taking care of it is extremely important." Robertson never mentioned climate change in the ad, only asking viewers to "get involved, it's the right thing to do." Still, a voiceover at the end of the spot asserted that "together we can solve the climate crisis," implying that Robertson now accepted climate change.

CBN's own story about the ad rejected such an interpretation, however, stating that Robertson was merely concerned to be a "good environmental steward" and "has never taken a position on the question of global warming."[24] Continuing in this ambiguous vein, on the March 31, 2008, broadcast of *The 700 Club*, Robertson discussed his participation in the campaign not in terms of climate change advocacy but because "I personally have always supported clean air, clean water and the reduction of acid rain. It's just common sense that we ought to be good stewards of the environment and do everything within our power to protect this fragile planet that we all live on."[25] In 2016 a representative from Robertson's public relations firm told me that Robertson had not taken a position on whether global warming was caused by humans but was "strongly opposed to how the whole concept of climate change has become heavily politicized to support a liberal agenda which only serves to harm American businesses."[26]

Despite this apparent non-position, as table 2 shows, Robertson's broadcasting network has leaned heavily toward skepticism on its broadcasts, promoting such views on *The 700 Club* and on *CBN News*. By Earth Day 2008, for example, just a month after Robertson's brush with activism, CBN gave ample coverage to skeptics in a news report titled "Is Global Warming Even Real?"[27] Edging further away from climate advocacy, on December 4, 2009, *The 700 Club* included a report on the 2009 United Nations Climate Change Conference in Copenhagen.[28] Helping to amplify the false Climategate story (see chapter 4), the report complained that "the media had mostly played down or ignored what many feel is a huge story"—Climategate—while also condemning climate-related alarmism. "Fear is a tremendous motivator. You either get motived by greed or you get motivated by fear, so now they're using fear," Robertson opined at the report's end.

A search of CBN.com's TV website identified fifty-eight stories discussing climate change between 2009 and 2015 (see table 2). Forty-six of them, or 79 percent, were skeptical, while just two were accepting of climate change; the remaining were either neutral or the position was unclear. This bias toward skepticism is significant, for according to its own accounting at least, CBN reaches 97 percent of U.S. television markets and draws one million daily viewers.[29]

In terms of its web presence, CBN.com—where all the skeptical stories for television as well as additional skeptical content produced for the web appear—also has a large footprint. An analysis conducted in 2017 found that it drew an average of 1.3 million visitors a month.[30] In fact, as of 2016 it was the third-highest-ranked Christian news website, behind only the *Christian Post* and *Christianity Today*.[31] CBN.com thus stands to substantially magnify the impact of skeptical stories appearing in CBN's television programs.

In sum, despite Robertson's reluctance to consistently advocate for a particular position on climate change, his network helped promote skepticism to a substantial audience.

CHUCK COLSON AND *BREAKPOINT*

Like Robertson, Colson had enormous potential to shape perceptions about climate change, in Colson's case via his popular radio feature *BreakPoint,* which is dedicated to "[cutting] through the fog of relativism and the news cycle with truth and compassion."[32] Partly on the basis of his previous political experience, Colson was regarded by many as one of the most influential evangelicals in politics.[33] Before becoming a Christian, Colson served as special counsel to President Nixon, but as a result of his involvement in the Watergate scandal, he was sent to prison in 1974. Before going to prison, he converted to Christianity, an experience that later inspired him to create a prison-based ministry called Prison Fellowship. He later founded *BreakPoint,* a four-minute radio feature that was broadcast (as of 2016) on twelve hundred outlets, giving Colson a significant platform for promoting climate skepticism.[34]

In 2009 and 2010, Colson devoted twenty-two of his *BreakPoint* commentaries to expressing skepticism about global warming (see table 2). The installment for October 2009, for example, publicized the skeptical documentary *Not Evil Just Wrong: The True Cost of Global Warming Hysteria.*[35] He described the film's content, summarizing a number of its skeptical talking points, and invited listeners to visit his website in order to obtain a copy. In December of that year, he released a video commentary criticizing the 2009 conference in Copenhagen. In the video he argued that the conference was not actually about climate change, but about

> control. It's about leverage to create the technocrat's vision of a better world. . . . This would give planners absolute control over the way people live their lives, the kind of control that hasn't been seen since the fall of Communism. Only this time it would be exercised in the name of polar bears and penguins, not the proletariat.[36]

The video—labeled "Two Minute Warning"—was clearly designed to persuade listeners that climate change advocates had a dangerous agenda. A day later, on *BreakPoint,* he railed against a court decision in Great Britain to treat belief in global warming as a form of religious belief in cases related to workplace discrimination. "Listen, folks," he concluded, ominously referring to the marginalization of Christianity, "in today's climate, the earth could soon enough take the place of that archaic idea of an ancient God of the Bible."[37] For anyone needing help connecting the dots between faith and climate skepticism, Colson's ministry provided the key ingredients.

A search of the *BreakPoint* archives suggests that the program addressed climate change only about half as often as it did issues central to the Christian Right—abortion and religious freedom—but, as of 2015, it had nevertheless devoted more coverage to the topic than it had to Islam or to the end times.[38] Furthermore, its coverage made no attempt to present both sides of the issue: as of 2016, the *BreakPoint* archives included fifty-four pieces of content that discussed climate change (ranging from 2006 to 2015), every single one of them skeptical.[39] In short, promoting climate skepticism was a significant part of the agenda of one of the most powerful evangelical broadcasters in America.

Colson also spread his views on climate change further via opinion pieces on highly trafficked websites. His 2010 radio opinion piece

"Global Warming and Ideology," which was broadcast on *BreakPoint*, argued that "global-warming hysteria has given license to an anti-human view that portrays people as a problem to be managed."[40] The transcript appeared not only on BreakPoint.org, but also on the website of the *Christian Post*, which by 2017 had an average of 1.1 million visitors a month;[41] as of 2016, it was the second-highest-ranked Christian news website, after ChristianityToday.com.[42] The transcript also appeared on Crosswalk.com, which is part of the fast-growing Salem Web Network, owned by the leading evangelical media group, Salem Media Group. As of 2007, the Salem Web Network was boasting six million unique visitors a month.[43] Two of Colson's other skepticism-promoting columns also ran on ChristianPost.com.[44] A number of Salem-owned radio stations that broadcast *BreakPoint* also posted transcripts of his commentaries about climate change to their websites. Hence, online media directed his radio message to an even wider audience. To be sure, Colson's views were not necessarily the only ones being presented on these websites, but their inclusion increased the likelihood that those who respected Colson would be influenced by his thinking on the issue.

D. JAMES KENNEDY, CORAL RIDGE MINISTRIES, AND ANSWERS IN GENESIS

As a Christian broadcaster who blended faith with politics, D. James Kennedy had much in common with the men described above. Not only was Kennedy the pastor of a ten-thousand-strong megachurch, but his Coral Ridge Ministries, which included radio and television ministries, reportedly reached 3.5 million people at the time of his death in 2007. Like Robertson and Colson, he argued that Christianity was under attack and urged Christians to "reclaim America for Jesus Christ."[45] Also like them, he used his media network to spread climate change skepticism.

Others have noted that Kennedy signed a number of the Cornwall Alliance's statements. But what is less well known is that Kennedy invited E. Calvin Beisner, the founder and national spokesman of the Cornwall Alliance, on his daily radio program, *Truths That Transform,* which was carried by seven hundred radio stations as of 2006.[46] In the transcripts, which were later published by Coral Ridge Ministries as the book *Overheated: A Reasoned Look at the Global Warming Debate,* Beisner argues that global warming will harm the world's poor by slowing their economic development and invites listeners to join the Cornwall Alliance. Although the interview is mostly a vehicle for Beisner's ideas, Kennedy concludes with a personal endorsement of climate skepticism, stating that "God has called us to care for our planet. In this day and age, this command includes having a balanced approach to caring for our natural environment. We are to take this responsibility seriously, without buying into the alarmist theories that have plagued this debate."[47] As of April 2007, in addition to marketing *Overheated,* the website of *Truths That Transform* was also offering the related title *The Politically Incorrect Guide to Global Warming and Environmentalism* for a donation of twenty dollars.[48]

Kennedy also promoted climate skepticism in his 2008 book *How Would Jesus Vote? A Christian Perspective on the Issues,* which he cowrote with Jerry Newcombe, who later became a senior producer and on-air host with D. James Kennedy Ministries. Urging their evangelical readers to disavow climate concern as incompatible with their faith, Kennedy and Newcombe argue that

> those committed to a biblical worldview understand that God made the world, and it does not hang in the balance. The world He made goes through warming and cooling periods that have nothing to do with human activities. It is simply human hubris when we declare that we are the cause of global warming (or cooling).[49]

After arguing that most evangelicals disagree with the ECI, the authors echo the combative framing I heard in Southern Baptist circles, stating that they are "disappointed that conservatives have largely abandoned the field to the liberals."[50] Much like many of the evangelicals I met in Georgia, Kennedy promoted the view that accepting climate change was incompatible with a Christian worldview. Newcombe would later promote climate skepticism on his own radio show, *Vocal Point* (see table 2).

After Kennedy's death, Coral Ridge Ministries extended its efforts further via a partnership with Answers in Genesis (AiG), perhaps the largest creationist organization in America. In 2008 the two organizations coproduced a skeptical documentary called *Global Warming: A Scientific and Biblical Exposé of Climate Change.* The documentary presents a mixture of conservative Christian leaders (including Beisner and Richard Land) and skeptical spokesmen, some of them scientists (although only two, John Christy and Roy Spencer, are climate scientists). Innovatively, *Global Warming* explicitly links anti-evolutionism with climate skepticism by including several spokesmen from creationist circles, such as Jason Lisle and Larry Vardiman, both of whom are or have been affiliated with the Dallas-based Institute for Creation Research and with AiG. Lisle and Vardiman advocate climate skepticism at length in the documentary, as does Dr. Jay Wile, a nuclear chemist who has written creation science–based textbooks for homeschoolers. According to Wile, it is belief in evolution that causes excessive anxiety about climate change:

> If you believe in evolution, then you believe random chance is the main architect, guided by natural selection. Random chance is a terrible designer, so anything it creates would be incredibly fragile. . . . If you believe instead that this is all the result of a very powerful, very intelligent designer, you assume He's put in feedback mechanisms and failsafes [sic] to keep the earth robust as it changes over time.[51]

Speakers with strong religious credentials like Beisner and Land are not the only ones to address religious concerns in the video. In one segment, the atmospheric scientist John Christy, who has gained notoriety in secular circles as a climate skeptic, reminds viewers that "the number one thing evangelical Christians should remember is, what is at the peak of the pyramid in terms of creation? And if you read Genesis especially, you'll see that human life is at the peak of the pyramid." Roy Spencer, Christy's colleague at the University of Alabama in Huntsville, also discusses his view of what environmentalism "from a biblical standpoint" should be. Their comments underline the entanglement between secular and evangelical climate change skepticism.

While it is difficult to determine how many people have seen the documentary, it would have received a boost from CBN's 2008 story about it, which presented it as a less agenda-driven alternative to *An Inconvenient Truth*.[52] Viewership aside, the documentary demonstrates Coral Ridge Ministries' willingness to devote resources to promoting skepticism as the Christian perspective on climate change.

Since Kennedy's death, others associated with Coral Ridge Ministries have continued his efforts to promote climate skepticism. John Rabe, who is credited with cowriting the script for the *Global Warming* documentary, has produced at least two video segments advocating climate skepticism for a D. James Kennedy Ministries online video series called "Learn2Discern."[53] Further spreading these views, on April 2014, D. James Kennedy Ministries repackaged an undated Kennedy sermon against idolatry as an anti-environmentalist response to Earth Day. After the sermon, which appeared to be at least a decade old, Kennedy's daughter transitioned to the topic of climate change, asking, "Do you realize there's a growing push to replace God with government in the environmentalist movement?"[54] Produced by Rabe and Newcombe, the segment aired many of the same skeptical arguments made by Beisner and his associates, while also promoting some of Beisner's writing on the topic. It was available to forty-two million households via its inclusion on the NRB Network (now NRBTV).[55]

As for AiG, its efforts went far beyond coproducing the *Global Warming* documentary. Most importantly in terms of audience, Ken Ham (president, CEO, and founder of AiG) addressed global warming twenty-eight times between 2008 and 2015 on his radio feature *Answers with Ken Ham* (see table 2). In 2008, for example, Ham used the feature to promote the *Global Warming* documentary, which he offered in return for a donation of any size. In advertisements that appended ten segments from October of that year, the announcer explicitly mentioned several times that climate change was "one of the hot topics during this election time." As we will see in chapter 8, the 2008 presidential election was a critical time for leaders in the Christian Right. Ham's support was thus well timed to protect the Christian Right's broader political interests. The features about global warming were presented alongside features addressing topics such as "Secular humanism—is it taking over society?" and "The American way—attack Christians?," further blending climate skepticism into the broader discourse of embattlement.[56] *Answers with Ken Ham* is only a minute long but has a substantial audience, being broadcast on nine hundred stations in forty-nine states plus Guam.[57]

Second, demonstrating AiG's insertion of global warming into its broader agenda of defending Christian values, the *Global Warming* DVD is sold in AiG's online store as part of a "Reclaiming America Pack," which also includes videos addressing abortion, stem cell research, and terrorism. A "pocket guide" version of the *Global Warming* DVD is sold there as well. Like the documentary, the pocket guide was promoted heavily on *Answers with Ken Ham*.[58]

In print media, the *New Answers Book* series, which addresses questions related to creation and evolution, includes chapters on global warming beginning in 2010 (Books 1 and 2, published in 2006 and 2008, respectively, do not mention global warming). The 2010 chapter, by Michael Oard (who also appears in the *Global Warming* DVD), accepts that the earth is warming but concludes that "man is likely responsible for only about 0.5°F warming—miniscule and likely impossible to mitigate."[59]

The 2013 chapter is equally skeptical of anthropogenic climate change, noting in conclusion that "it is humbling to remember that when God was judging the earth with a global flood that He was creating inexpensive fuel sources for future generations."[60] Demonstrating the interconnectedness of various media, in 2015 Ham promoted this book, including its take on climate change, on a segment of the radio program *The Christian Worldview,* which airs on 225 stations around the country.[61]

Issues of *Answers* magazine, which claims two hundred thousand readers, discussed global warming in 2006, 2008, 2010, and 2016. In addition, AnswersinGenesis.org offers additional content promoting climate skepticism, with a total of forty-two stories posted between 2005 and 2016, some of them penned by Ham himself. As of August 2017, the site was expected to have 740,000 average monthly users.[62]

Like Robertson, Ham is not affiliated with the Cornwall Alliance, nor is he widely known outside of evangelical circles as a leader in the Christian Right. Yet he has appeared on many of the right-of-center Christian talk radio shows I reviewed for this analysis, and, perhaps more importantly, he is committed to the idea shared by many in the Christian Right that liberals and secular humanists are destroying America's biblical foundations.[63] While his take on climate change differs from that offered by others in the Christian Right (he argues that the climate is still settling down after the flood described in Genesis), his work in this area has been highly synergistic with their efforts—as illustrated by his collaboration with Coral Ridge Ministries.[64]

JAMES DOBSON, FOCUS ON THE FAMILY, AND THE FAMILY RESEARCH COUNCIL

James Dobson has been less directly involved in the climate skepticism campaign than those just described, but investigating the efforts of his associates revealed significant activity. Dobson began his career on the faculty of the University of Southern California School of Medicine, where his interest in family issues led him, in 1977, to found the parachurch organization Focus on the Family (which he ran until 2003) and to launch a radio program of the same name. Enormously popular, the broadcast boasted 3.4 million weekly listeners on two thousand stations by the time climate change became a hot topic in evangelical circles.[65] However, despite early efforts to oppose the ECI (described in chapter 8), Dobson soon distanced himself from the skepticism campaign. Instead, a lobbying and public policy–focused offshoot of Dobson's Focus on the

Family, known as the Family Research Council (FRC), has done most of the legwork. This delegation makes sense in light of the FRC's origins as a way for Focus on the Family to pursue a political agenda without Dobson himself being tarnished with an overly partisan image.[66] Dobson created the FRC in 1983; by the end of the 1990s, with a $14 million budget and 120 staffers, the FRC had become the most powerful Christian Right group in Washington.[67] It would use this power to amplify climate skepticism within conservative evangelical circles.

In 2006, Dobson had a newsletter that reportedly went out to 2.5 million people, but he did not use it as a vehicle to promote skepticism.[68] He also distanced himself from the Cornwall Alliance, despite sharing its concerns. While Dobson had signed its founding document in 2000, his name did not appear on the Cornwall Alliance's initial response to the ECI.[69] However, the current president of the FRC, Tony Perkins, was listed as a signatory. Perkins, who has served as president of the FRC since 2003, also signed documents that the Cornwall Alliance released in 2009 and 2012.[70] Dobson's name did not appear on either document, further indicating that the FRC had taken the lead in efforts to oppose evangelical climate activism.

Despite its non-environmental mission of "advanc[ing] faith, family and freedom in public policy and the culture from a Christian worldview," the FRC has been closely involved in promoting climate skepticism since the early days of the evangelical debate on the topic.[71] In 2006, a senior fellow with the FRC penned an opinion piece in the conservative newsweekly *Human Events* arguing that "Evangelicals Should Not Be Fooled by Global-Warming Hysterics."[72] In 2007 the FRC hosted an informal debate between the Cornwall Alliance and the EEN.[73] The next year it was one of just five organizations (besides the Cornwall Alliance itself) to endorse the Cornwall Alliance's "We Get It!" campaign. In a creative expansion of the pro-family agenda, it linked the issue to its mission statement by stating that the increased food and energy costs that would be incurred by "an environmental threat that is at best speculative" threatened the family.[74] Perkins also appeared in the Cornwall Alliance's anti-environmental documentary film *Resisting the Green Dragon* (2010). Further demonstrating the close working relationship between the FRC and the Cornwall Alliance, on Earth Day 2010, Beisner gave a lecture at the FRC's media center entitled "How a Climate Change Treaty Threatens You, Our Nation and Your Church."[75] The lecture was later made available to the wider public via webcast.

Beyond official partnerships, Perkins has also harnessed his organization's media apparatus to disseminate climate skepticism directly to laypeople. His radio program *Washington Watch* devoted seven segments to climate change in 2014 and twelve in 2015 (see table 2). Given that it is a daily program, this frequency of coverage is modest. Yet the message for anyone listening was unmistakable. For example, the November 18, 2015, program began like this: "'You load people down with burdens they can hardly carry, and you yourselves will not lift one finger to help them.' Hello, this is Tony Perkins with the Family Research Council in Washington. Jesus' words to the Pharisees could just as easily be applied to the elites driving the climate change debate today."[76] As this phrasing suggests, the broadcasts related to climate change adopted a strongly skeptical perspective, presenting it as the only sane, reasonable position on the issue. On weekends (as of 2016) the program aired on three hundred radio stations around the country.[77] Perkins and other FRC associates have also written a number of posts on the FRC's blog mocking, denouncing, or ridiculing climate change advocates.[78]

Perkins also addressed climate change in his book *Personal Faith, Public Policy,* coauthored with the conservative Pentecostal pastor Harry Jackson Jr. Echoing what I heard in my focus groups, the authors accept that the earth is warming but question the cause and criticize "global warming alarmists," who "have their 'solutions' waiting in the wings, and these solutions mostly involve greater government intervention."[79] In a summation that echoes the sentiments I heard in my focus groups, Perkins and Jackson write that

> [deceived people] put their faith in something other than God. And this is exactly what environmental alarmists ask us to do. The language they use and the policies they promote are humanistic. How many times have you heard them appeal to others to help save the planet? The whole premise of the statement presupposes that mankind is ultimately in charge of the fate of our planet. It springs from the same idea that we can save ourselves and that we don't need the atoning work of Christ on the cross. If the problems of pollution, the environment, and global warming are man-made, their logic goes, then the solution can be man-made too. We don't need God. If undue focus on the environment is indeed a type of spiritual deception, then its proponents are conditioning people to look to government and to the powers of man—not God—to save them.[80]

With statements such as these, echoed and amplified via radio and online, it is not difficult to imagine how this particular theological take on climate change became so dominant among traditionalist evangelicals.

It is worth adding that Perkins and Jackson also hinted that climate change might be a sign of the end times. Consistent with their emphasis on this-worldly engagement, however, they underlined that "this declaration is not meant to be fatalistic." Instead, following the pattern discussed throughout this book, they concluded that "it is time for the church to add their [sic] voice to the discussion of environmental issues without adopting the at times godless tactics, agenda, and philosophy of the more extreme alarmists."[81]

The FRC also lent its public policy expertise to the cause of climate skepticism. In 2008—a critical year for backlash against the ECI (see chapter 8)—the FRC released a document called "25 Pro-Family Policy Goals for the Nation," which was intended to "serve as a blueprint for how those we elect can promote and protect the family and its values." Even though it was focused primarily on family issues, it also included proposals related to climate change, including advocating for delay on some policy measures until "catastrophic climate change . . . can be ascertained with replicable scientific evidence" and protecting climate skepticism as "academic freedom"—a clear demonstration that climate change skepticism had been added to the organization's agenda.[82]

The FRC also intermittently releases background materials on nominees for public office, highlighting their positions on certain issues, on its blog. Usually these issues are related to the family (e.g., stem cell research or abortion), but reviewing these materials reveals that in 2009 the FRC also created a number of "backgrounders" on individuals nominated for environmentally consequential positions. These backgrounders focused specifically on the candidates' views on climate change, listing quotations that clarified their positions. For example, for Jane Lubchenco, who in 2009 was nominated to be administrator of the National Oceanic and Atmospheric Administration, the backgrounder listed fourteen public statements she had made about climate change.[83] The backgrounders further indicated the extent to which the FRC now considered climate change to be part of its agenda: acceptance of climate change was now cast in the same negative light as a permissive stance on abortion.

Further demonstrating the FRC's efforts to lobby in favor of climate change skepticism, FRC Action, the FRC's lobbying political action committee, included climate change in materials it created to assist local churches in holding news-ready town hall meetings. Labeled a "Town-hall Kit," the materials included a template letter that church pastors could customize and send to their congressional representatives, inviting them to participate. By way of enticement, the letter highlighted

the opportunity to address a large number of constituents interested in hearing directly from you on important issues such as the law on open homosexuality in the military, repealing the health care law, and a possible lame-
duck session that could include votes on global warming, more taxpayer
funding of abortion, and illegal immigration.[84]

Hence, global warming clearly belonged in the stable of issues that were
occupying the FRC's attention at the time.

Finally, the FRC hosts the annual Values Voter Summit, where (as the
next chapter relates) Senator Inhofe would urge politically active evangelicals to get involved in promoting climate skepticism. Given the overall
picture of the FRC's involvement in climate change skepticism, Inhofe's
decision to talk about climate change at such an event made perfect sense.
Indeed, Inhofe followed his own advice by warning the faithful about
global warming on several occasions on Perkins's radio program.[85]

JERRY FALWELL AND *THE OLD-TIME GOSPEL HOUR*

The efforts of Jerry Falwell, the last of the Big Four leaders whose work
I examined, were significant but cut short by his death in 2007. Falwell
was a Baptist minister who became nationally known as a televangelist
via his radio and TV program *The Old-Time Gospel Hour*.[86] Broadcasting his sermons from Thomas Road Baptist Church in Lynchburg, Vir-
ginia, the show attracted 1.2 million regular viewers by 1980.[87] Beginning in the late 1970s, Falwell also became politically active. Though he
had already been fighting against secular humanism for some time, it
was Francis Schaeffer who convinced him to take his fight into the political realm.[88] He did so by founding, in 1979, the Moral Majority, which
(with the help of two other groups) registered about two million new
voters in time for the 1980 election. While relatively small in terms of
its constituency, the Moral Majority played an important role in turning evangelicals into a political force.[89]

Falwell was an early climate skeptic. In a testy 2002 exchange with
Ron Sider over the EEN's "What Would Jesus Drive?" campaign (see
chapter 8), Falwell asserted that global warming was a myth and pug-
naciously urged viewers to "go out and buy an SUV today."[90] Although
his opinions about climate change had clearly gelled early in the decade,
it was not until after the ECI was released that he began to advocate
climate skepticism more systematically. In February 2006, he penned an
opinion piece on the Christian online news website WND.com, where
he was a commentator, arguing that "global warming is an unproven

phenomenon and may actually just be junk science being passed off as fact."[91] While he did not mention the Cornwall Alliance, his letter mirrored its arguments to the point of quoting Paul Driessen, who has publicly represented the Cornwall Alliance as well as coauthored several of the documents it produced urging skepticism about climate change, and Alan Wisdom, who was at the time on the advisory board of the Cornwall Alliance's predecessor, the Interfaith Stewardship Alliance.[92] A second piece, written a few weeks later, sought to quash support for the ECI by announcing that it had accepted funds from the Hewlett Foundation, an organization that (it implied) should be viewed with suspicion because it also supported Planned Parenthood, an abortion provider.[93] In November he published a third piece on WND.com promoting climate skepticism.[94]

With an average of 1.3 million visitors a month, WND.com has a significant audience.[95] In fact, it is the second-highest-ranked religion website (behind only BibleGateway.com), coming in slightly ahead of secular sites like MSNBC.com, PBS.org, and RollingStone.com in terms of average numbers of daily visitors and pageviews per month.[96] Hence, Falwell lent his name and reputation to the cause of promoting climate skepticism in a venue that maintained a high profile within the evangelical world.

Falwell also delivered at least three sermons on climate change at his twenty-one-thousand-member church.[97] Like Falwell's other sermons, these would have been broadcast locally on the Liberty University–affiliated radio station (WRVL) and nationally via Falwell's syndicated radio and television broadcasts.[98] Although its own actual viewership was likely far lower, the Liberty Channel, on which Falwell's sermons aired, was available via Sky Angel, a satellite streaming service that claimed to reach twelve million households in 2007.[99] Demonstrating the considerable crossover with the media empires described above, in addition to broadcasting Thomas Road's weekly service twice daily, at that time Falwell's radio station also carried *BreakPoint, Focus on the Family, Answers in Genesis,* and *Truths That Transform.*

In a sermon delivered on February 25, 2007—and announced in a WND.com column—Falwell attacked climate change as a tool concocted by Satan to take Christians' attention away from more important issues.[100] Like the articles, it was styled as a direct attack against the ECI, whose participants he labeled "first-class nuts."[101] Like the statements of other skeptical leaders in the Christian Right (and echoing his earlier thoughts on the topic), Falwell's sermon contained a mixture of secular skepticism supported by theology and reinforced with the language of embattlement. After quoting a range of skeptical talking points, he described global warming as "the greatest deception in the history of science." Like Colson, he suggested that climate advocates were allied with evangelicals' enemies, the secular elites: "liberal politicians, radical environmentalists, liberal clergy, Hollywood and pseudo-scientists." Turning to theology, he reassured his audience that, as promised in Genesis 8:22, the earth's climate would be stable until the end times. Finally, echoing a common sentiment among my interviewees, he argued for a moderate, "reasonable" environmentalism that excluded concern about global warming:

> Every Christian ought to be an environmentalist of [a] reasonable sort. . . . We should certainly pick up trash. We ought to beautify the earth as best we can. We ought to keep the streams clean. But we shouldn't be hugging trees and worshipping the creation more than we worship the Creator, and that is what global warming is all about.[102]

In addition to delivering the sermon, Falwell also posted "A Skeptic's Guide to Debunking Global Warming Alarmism" on the website for Thomas Road Baptist Church. The link to the document, a seventy-page booklet assembled by Senator Inhofe's staff in late 2006, appeared prominently on the home page. Falwell reminded his audience (in his church and across America) no fewer than three times to access the information.[103]

Falwell was not necessarily as popular with younger evangelicals during the last years of his life, but among those who viewed him positively, his endorsement of climate skepticism would have been significant.[104]

While climate change is no longer a central topic at Falwell's church, there are a few hints that his skeptical legacy has lived on through Jerry Falwell Jr. In 2007, the younger Falwell became president of Liberty University (which was founded by his father in 1971). In 2009, Falwell Jr. argued that the graduating class "needs to be prepared to recognize the fraudulent use of science [including] the use of global warming theories to advance environmental extremism."[105] The next year, Falwell invited Lord Christopher Monckton, a British public speaker who is well known for his advocacy of climate denial, to speak at the university's convocation. Falwell introduced Monckton with a warning about the "false fear of climate change." Monckton elaborated on the theme by speaking about the "fallacies" of global warming and disparaging Al Gore's film *An Inconvenient Truth* as a "mawkish, sci-fi, comedy horror movie dreamt up by a PR guy."[106] In a phrasing that nicely illustrated how climate skepticism had been woven into the worldview of American conservative Christians, the university press release about the speech praised Monckton for calling students "to once again challenge majority opinion, media content and so-called scientific facts."[107] With thirteen thousand attendees, Liberty University's convocations are, according to the university's own description of the events, the largest weekly gathering of college students in the world.[108]

BEYOND THE BIG FOUR: FURTHER EFFORTS TO PROMOTE CLIMATE SKEPTICISM

While the Big Four stand out in terms of reach, a number of others associated with the Cornwall Alliance have also promoted climate skepticism. Examining their efforts reveals how the significant media resources controlled by evangelicals who opposed action on climate change enabled them to amplify their message.

David Barton

Probably the most influential evangelical leader to help promote climate skepticism beyond the Big Four is David Barton, best known for his work promoting the ideas that America was founded as a Christian nation and that Christianity should continue to play a prominent role in government (many scholars regard Barton's historical writing as deeply flawed).[109] He was listed as one of *Time* magazine's twenty-five most influential evangelicals in 2005, and his books on the role of Christianity in American culture have been best sellers.[110]

Barton's primary vehicle for promoting skepticism was his daily radio show, *WallBuilders Live!*, which addressed climate change twelve times between 2008 and 2015 (see table 2). Four episodes he devoted to "Science, the Bible and Global Warming" in 2008 were later packaged as a CD for sale through his online store, increasing their reach.[111] The program is broadcast fairly widely, being carried on almost two hundred radio stations nationwide.[112] In a 2010 appearance on Glenn Beck's Fox News program with the Cornwall Alliance's national spokesman, Barton further argued that "there's climate change, but it's not man-caused."[113] In 2008 Barton's organization, known as WallBuilders, distributed a voting guide that devoted several paragraphs to criticizing the ECI for trying to redefine "moral issues."[114]

Wayne Grudem

Wayne Grudem's contribution to the evangelical climate skepticism campaign is important because of his credentials and influence within evangelical circles. Grudem's background—a bachelor's degree in economics from Harvard and a PhD in New Testament from the University of Cambridge—mark him as a serious intellectual. He has gained further renown in evangelical circles via high-profile publications, including serving as general editor of the *English Standard Version Study Bible* and author of the popular seminary textbook *Systematic Theology* (a book that was recommended to me while I was in the field). In addition to publishing extensively on topics ranging from evangelical feminism to "the moral goodness of business," Grudem has also made a point of advocating that Christians be involved in the nation's political life.[115] Grudem speaks with authority, which adds great weight to his views regarding climate change.

In terms of disseminating climate skepticism, Grudem does not have the reach of radio hosts like Barton, but for anyone interested in his views regarding climate change, his website offers links to an audio lecture on climate change that was recorded as part of a lecture series for an adult Bible class at a megachurch in Arizona.[116] Like others discussed above, Grudem combines skeptical talking points with theological objections to climate change. As for theological objections, he has echoed themes I heard in the field, including that "the underlying cause of fears of dangerous global warming might not be science, but rejection of a belief in God—his goodness in creating the earth, and his goodness in sustaining it."[117]

The material Grudem covers in his lectures is also available in greater detail in a chapter on "The Environment" in his book *Politics according to the Bible* (2010). Notably, in the acknowledgments to this text, he describes Beisner as "the world's leading expert on a Christian perspective on uses of the environment."[118] The acknowledgments also relate that Grudem allowed Beisner to write the first draft of the section on global warming.[119]

In 2015, Grudem joined with other Cornwall Alliance associates to criticize the 2015 Paris climate talks in an opinion piece published on FoxNews.com.[120] The piece articulated an economic rather than faith-based critique of attempts to address climate change, but the inclusion of Grudem's name on the byline would have lent weight to that opinion among those who respected Grudem's reputation as a biblical scholar (the lead author was also a biblical scholar and Cornwall Alliance supporter).[121]

Hence, with Beisner's help, Grudem has further spread the message that skepticism is the biblical view on climate change, integrating it into his larger project of encouraging evangelicals to become politically engaged.

Janet Parshall

Janet Parshall is a third Cornwall Alliance associate who has promoted climate skepticism widely, in her case primarily through the radio program *In the Market with Janet Parshall.*[122] From 2010 to 2015, she devoted twenty-three segments to promoting climate change skepticism (see table 2).[123] The show has an impressive reach: as of 2016 it was broadcast on 372 stations in the United States.[124]

Other Purveyors of Climate Skepticism

A number of others (some but not all of them supporters of the Cornwall Alliance's initiatives) have contributed to the campaign. For example, John Sumser's three-month audit of *The Janet Mefferd Show* and Bryan Fischer's *Focal Point* in the spring of 2014 revealed that Mefferd's and Fischer's shows both portrayed climate change not just as "a liberal power-grab to impose more government regulations" (as is common among secular conservatives) but as "yet another attack by the culture on Christianity."[125] Until late 2016, Mefferd's show was broadcast in three hundred radio markets; as of 2014, Fischer's aired on over two hundred stations.[126]

Kevin Swanson's show *Generations Radio,* which describes itself as "the largest homeschooling and Biblical worldview program," included nineteen programs that discussed climate change from a skeptical perspective between 2006 and 2015 (see table 2).[127] The show has a smaller audience than those discussed above but claims to have had more than one million downloads of its programs.[128] *The Christian Worldview,* a weekly radio and online ministry that, as of March 2016, aired on 225 stations around the country, devoted six programs to climate change or environmentalism between 2008 and 2015.[129] Finally, Kevin Boling's radio program *Knowing the Truth,* which airs in North and South Carolina, Tennessee, and Georgia, included twenty-four shows about climate change or environmentalism from 2009 to 2015, thirteen of them with Cornwall Alliance associates (see table 2).[130]

Ted Baehr, president of the Christian Film and Television Commission, also contributed to the climate skepticism campaign via his website

Movieguide.org. While Moveguide.org ostensibly focuses on "redeem[ing] the values of the entertainment industry, according to biblical principles,"[131] the site has also included climate-related stories that are not about films but simply report the latest skeptical news about climate change.[132] According to my analysis, MovieGuide.org published forty-one items (movie reviews, opinion pieces, and news items) that referenced climate change between 2006 and 2015. Twenty-three presented skeptical talking points and an additional fifteen were critical of climate change without mentioning these talking points (i.e., criticizing a movie for promoting global warming hysteria without unpacking the claim). The remaining three were neutral or unclear; not a single story presented information reflecting the scientific consensus.

While the climate skepticism campaign has not relied heavily on print media, *World* magazine is an important exception. With an average circulation of one hundred thousand, its readership is just slightly smaller than that of *Christianity Today,* making it an influential publication, especially among conservative evangelicals, who are its primary audience. Its online avatar, World.wng.org (formerly Worldmag.com), boasted slightly more unique visitors per month than did ChristianityToday.com as of 2016, further speaking to its influence.[133]

Joel Belz, the founder and publisher of *World* (now retired), and Marvin Olasky, the magazine's editor in chief, have both been involved with the Cornwall Alliance, Olasky since 2000 and Belz since at least 2006.[134] A content analysis of *World* magazine's coverage of environmental issues found that during the pivotal period of 2004 to 2010 the magazine "grew increasingly loud in its arguments that evangelicals should not embrace environmental political issues" and "saw environmentalism as just another component of liberal politics and antithetical to evangelicalism."[135] *World* magazine has thus lent its influence in conservative evangelical circles to the cause of climate skepticism.

Guided and supported by the Cornwall Alliance, leaders and pundits associated with the Christian Right expertly used the evangelical mass media to encourage climate skepticism among evangelicals around the country, delivering messages whose tone and content bore a striking resemblance to those that I and other researchers heard in the field. With so many voices discouraging concern, often on explicitly religious grounds, it is unsurprising that climate skepticism came to be entangled with faith for many American evangelicals. Indeed, in the general region where I did my field research, at least eight programs promoting climate skepticism could be heard on local radio stations: *Answers in Genesis, The Janet Mefferd Show, Knowing the Truth, BreakPoint, Focal Point, WallBuilders Live!, In the Market,* and *Washington Watch.* And, of course, CBN was available as well via cable, satellite, or the Internet.[136]

I did not make a regular habit of asking my informants about their sources of information, but the consistency between the messages I encountered in the evangelical mass media and what I encountered in person seems well beyond what could be attributed to coincidence.[137]

While the details of the campaign—comprised of so many different individuals and operating across multiple media and media platforms—can be overwhelming, the overall picture suggests that the evangelical mass media campaign to promote climate skepticism played an important role in establishing skepticism among traditionalist evangelicals as an expression—in fact, a defense—of faith. Although many factors shape Americans'—and, especially, politically conservative Americans'—attitudes toward climate change, the campaign also deserves attention for its potential contribution to rising rates of skepticism among white evangelicals over the period. In July 2006, before the campaign began, only 25 percent of white evangelicals disbelieved that the climate was changing. By the fall of 2014, that number had grown to 39 percent.[138] Similarly, in 2006 a strong majority, almost seven in ten, thought global warming was a serious problem. In 2014, that number had fallen to less than half.[139]

The final question was why. Why had certain conservative evangelical leaders suddenly decided it was necessary to devote their time, energy, and resources to advocating a particular position on an issue so disconnected from their central areas of concern? In chapter 8, I propose an answer.

Awakening the Sleeping Giant

Quantitative studies of evangelicals' skeptical attitudes toward climate change have tended to attribute these attitudes to static, ahistorical, or slowly changing factors such as theological beliefs, political views, implicit conceptions of nature, or the like.[1] If my analysis is correct, however, leaders and pundits associated with the Christian Right instead actively coached these views, much in the same way that evangelical leaders in the 1970s worked to convince the laity that abortion was a moral calamity that needed to be addressed by political engagement.[2] But why?

Understanding motives seems necessary, because it would take a significant amount of effort and resources to shape the views of participants in a movement as large and decentralized as American evangelicalism, and because climate change does not seem, on its face, to be an issue that would naturally become a point of passion for a group that has more typically focused on issues such as sexual morality and religious liberty. How did climate change come to be of such intense interest? Furthermore, survey evidence suggests that evangelicals were already relatively skeptical about climate change in the early 2000s.[3] Why would leaders in the Christian Right devote resources to linking skepticism to faith when evangelicals already tended to be more skeptical than the general populace?

This chapter represents my answer. It is one that draws on publicly available materials; as such it does not purport to reveal the inner motives of individual actors but rather demonstrates that in the mid- to late 2000s, a great deal was at stake for leaders in the Christian Right in the question of whether American evangelicals would join the fight for climate action.

To make this case, I draw heavily on the excellent scholarship on evangelical environmental engagement during the first decade-and-a-half of the twenty-first century, especially the work of Laurel Kearns, Katharine Wilkinson, and the collaborators Lydia Bean and Steve Teles.[4] Because my goal is to explain the motivation for the Christian Right's campaign, which has not been discussed in previous research, I supplement these accounts with my own analysis of published materials produced by involved parties, and of media coverage of the greening of evangelicalism, which I deem a key factor motivating the campaign.

Since much of the history of evangelical environmentalism has been told elsewhere, in what follows I focus on describing the historical factors that motivated the campaign and enabled its success. Readers unfamiliar with the basic timeline of events are advised to consult the sources just mentioned. I begin with the long view, before delving into the events of the mid-2000s, when climate change became a topic of heated debate in evangelical circles.

CREATING THE CONDITIONS FOR FAITH-BASED ANTI-ENVIRONMENTALISM TO FLOURISH

In the long view, the Christian Right's climate change campaign was successful because it built on a decades-long alliance between white evangelicals and political and economic conservatives in the Republican Party. The roots of the evangelical-Republican alliance can be traced back deep into the twentieth century, but for the present purpose it makes sense to begin in the late 1970s, drawing connections between the histories told separately in chapters 1 and 4 about the anti-environmental backlash movement and the political mobilization of evangelicals.

In the late 1970s, a group of young conservative political activists dissatisfied with the direction of the Republican Party began promoting a new vision that favored laissez-faire economic conservatism, anticommunist foreign policy, and traditional values.[5] Known as the New Right, these activists built their coalition by seeking out the support of evangelicals, believing them to be, in the words of Richard Viguerie, "the next real major area of growth for the conservative ideology."[6] Their efforts to convince evangelicals to mobilize politically were aided by all the factors mentioned in chapter 4 as contributing to evangelicals' sense of embattlement with secular culture.[7] It was the genius of leaders in the New Right to convince these evangelicals—who came to oppose big government on the grounds that it threatened religious freedom and traditional values—to join with neoliberal "market fundamentalists" who opposed big government as a threat to "economic freedom" or free enterprise.[8]

As described in chapter 1, the new small-government coalition, which came into power with the election of Ronald Reagan, threatened the environmental movement because environmentalists had relied heavily on the federal government's enforcement to achieve its goals in the 1960s and '70s.[9] But the threat to environmentalists' interests did not come simply from economic conservatives; examining the coalition's origins reveals an important connection between anti-environmentalism and the political mobilization of evangelicals. As mentioned in chapter 1, before his cabinet appointment, James Watt had worked for the MSLF, an organization that Joseph Coors, the brewery magnate, had created in order to oppose intrusive environmental regulations.[10] Coors not only donated to anti-environmentalist causes, but also funded two New Right organizations that were instrumental in bringing evangelicals into conservative politics during this period: the Heritage Foundation and the Committee for the Survival of a Free Congress.[11] In fact, demonstrating significant crossover between the two groups, Watt had worked with the Heritage Foundation in the summer of 1980, before he was tapped to head the Department of Interior.[12] Nor was Coors an insignificant player: according to Viguerie, his financial backing was critical to the New Right's success.[13]

To some extent, then, the mobilization of evangelical Christians and anti-environmentalist forces were not a coincidence but rather separate entries in the same political playbook. Reduced environmental oversight was one goal of business interests that supported the New Right's small-government agenda; it was made politically feasible by the mobilization of evangelical Christian voters. And the consequences were long lasting: Coors's Heritage Foundation would go on to become the second-best-funded think tank involved in the climate change countermovement, receiving $76.4 million between 2003 and 2010.[14] As described below, it would also collaborate closely with evangelical anti-environmentalist activists when the controversy over the ECI erupted.

This background casts the uproar that Watt caused during his time as secretary of the Interior in a new light. Environmentalists of the early 1980s were right to suspect that Watt's religious beliefs were linked to his philosophy of natural resource management, but they had misidentified the connection. The most salient connection was not his end-time beliefs but his desire for small government—a desire that was closely tied to his sense of belonging to a faith that was under attack. More than an otherworldly apocalyptic, that is, James Watt fit the profile of an embattled evangelical. Indeed, perhaps more than any other figure of the period, he demonstrated the ways in which the New Right–Christian Right alliance would impinge on environmental politics.

Watt began his program of administering the nation's public lands with what he described as a "bias for private enterprise."[15] But for Watt, small government, free enterprise, the development of natural resources, and defending the Christian faith against attack were all of a piece. In a book published shortly after he resigned from his cabinet post, *The Courage of a Conservative,* Watt explained his views at length, arguing that it was the combination of both political and religious freedom that had originally "made America great." In his view, however, things had since gone awry, and "serious pruning" of the federal government was necessary in order "to recapture the original American idea."[16] In the realm of natural resource management, this meant that he believed that (as an analyst later summarized) "misdirected implementation of environmental protection laws had smothered opportunity and stifled productivity."[17] In Watt's view, these problems could be solved by encouraging development over preservation and—following the New Right philosophy—by reducing the size of the federal government.

In *The Courage of a Conservative,* Watt also made clear that his desire for a pruned federal government was connected to his religious outlook. In a chapter decrying "A Loss of Absolutes," he condemns those in the "liberal Establishment" who "seek to centralize government power in order to exercise social and economic control over the 'ordinary' individual."[18] Deploying the language of embattlement, he warns that "the elite Left seeks a humanistic, nonreligious society. They attack the integrity of religious institutions and leaders whom they regard as insincere or dangerous."[19] The only way out of this "intolerable struggle," as he views it, is to "stop the erosion of religious liberties and to recapture the values that have been at the very heart of the American spirit for a century and a half."[20] For him, this can best be accomplished by limiting the federal government's powers.

The connections between the growth of anti-environmentalism and the political mobilization of evangelicals speaks to a broader point: Watt has historically been regarded as revealing evangelical Christianity's inherent anti-environmentalism. But in fact he represented the moment at which leaders in the Christian Right, relying heavily on their coalition partners, began to help transform anti-environmentalism from a latent theme in Christianity to one that was relatively central to certain white evangelicals' faith and identity.

The potential for anti-environmentalism had long lurked in factors such as evangelicals' heavy emphasis on salvation and saving souls, suspicions of the environmental movement as harboring pagan and New Age sentiments, and apocalyptic speculation. But in the 1960s and '70s, these potentially anti-environmentalist impulses competed with a growing interest in environmental protection in evangelical circles.[21] It was in the 1980s that things began to change: as evangelical identity became tied to political and economic conservatism, it also became increasingly tied to anti-environmentalism.

Previous research has shown that New Right strategists encouraged white evangelicals to adopt their business-friendly worldview.[22] In the course of doing so, they also gave newly prominent leaders in the Christian Right ample opportunity to put a religious spin on economically conservative critiques of the environmental movement. In 1991, for example, Pat Robertson, who had run for president on the Republican ticket just two years before, warned against accepting "government as God" while also lamenting that "all over this country, children are being indoctrinated as world citizens, with reverence for the earth, the environment, the animals, and for people of all ethnic, religious, and sexual orientations."[23] Gary Bauer, a conservative political operative who got his start in Young Americans for Freedom and later headed the Family Research Council, also attacks environmentalism in a 1996 book published by Focus on the Family.[24] Chuck Colson argues in his book *A Dance with Deception* that Christians should be wary of the environmental movement because it is a form of humanism, arrogantly asserting that humans can solve environmental problems. "We are *not* God and we cannot control all the variables," he writes, foreshadowing his argument on climate change. The solution is to be found in the Bible's teachings about environmental concern, which in his view show that "God intended us to develop the potential in creation through industry and technology."[25] Hence, as leaders in the Christian Right were integrating economic and political conservatism into the evangelical identity, they were also—in a less consistent but nonetheless meaningful fashion—integrating skepticism of environmentalism.

While helping to nurture free-market-based suspicions of environmentalism among the laity, the New Right–Christian Right alliance also bore institutional fruit that would later be critical to the Christian Right's campaign to promote climate change skepticism. Most importantly, the Cornwall Alliance was created by a pro-free-market think tank known as the Acton Institute, which, in a blend of New Right and Religious Right ideas, aimed to demonstrate "the compatibility of faith, liberty, and free economic activity."[26] The economic ideology favored by New Right activists can be seen clearly in the writings of the Cornwall Alliance's leader. Indeed, according to one historian, his environmental vision was so indebted to neoliberal economic theory that it was "little different than free market economics given a Christian veneer."[27]

From a tradition containing the seeds of a potent environmental praxis—illustrated by Francis Schaeffer's argument that Christianity was uniquely suited to promoting environmental concern (see chapter 4)— a very different fruit emerged. Yet while the New Right–Christian Right alliance set the stage for evangelical anti-environmentalism to flourish, it would take a specific series of events to push leaders in the Christian Right from a low-investment strategy of intermittently critiquing the environmental movement to a high-investment strategy of actively promoting their favored view on a particular environmental issue among the evangelical laity.

FROM PASSIVE TO ACTIVE ENGAGEMENT

The key events that would ultimately push leaders and pundits associated with the Christian Right to initiate a campaign to disseminate climate skepticism were the ECI and the SBECI. The ECI's launch in February 2006 created confusion about where evangelicals at large stood on a consequential political issue in a way that was threatening to the power of leaders in the Christian Right, while the SBECI's launch in March 2008 made it clear that the push for climate action would not dissipate of its own accord. A closer look at the Christian Right's place within the larger evangelical tradition, as well as the circumstances under which climate change came to be a topic of debate in evangelical circles in the 2000s, will clarify the substance of this threat.

The Christian Right's Weakening Hold and the Rise of the "New Evangelicals"

Secular observers often imagine that pastors are powerful men who can lead their congregations in whatever direction they wish. In talking with pastors, however, it was clear to me that their authority was much weaker than is commonly supposed. The constant risk that church members would go elsewhere if they did not like what they were hearing left pastors with surprisingly little room to advance unpopular opinions.

Perhaps even more surprisingly, the situation among politicized evangelical elites is similar. In contrast to a hierarchical tradition like Catholicism, where authority is granted from the top down, in evangelicalism authority comes largely from the ability to attract followers (while maintaining a reputation for orthodoxy).[28] The Christian Right's power seems impressive from the outside, but its leaders are powerful because the message they send continues to resonate with supporters. If it stops resonating, or if a new vision captures the evangelical imagination, that work can be undone. It was exactly this vulnerability that evangelical environmentalists moved to exploit as they worked in the early 2000s to nudge climate change to the center of the evangelical tradition.[29]

Conveniently for them, the time was ripe for an insurgency, for around the same time a number of popular evangelical leaders had begun to challenge the Christian Right's control over the evangelical agenda. Not coincidentally, several of these men became prominent supporters of the ECI—not always because they were passionate about climate change, but because embracing creation care embodied their vision for a broader, more generous style of evangelical social engagement.

Joel Hunter, Rick Warren, Brian McLaren, and David Gushee are among the most prominent of these leaders. Hunter, a megachurch pastor and best-selling author, forcefully makes a case against the Christian Right's leadership style in his 2006 book *Right Wing, Wrong Bird: Why the Tactics of the Religious Right Won't Fly with Most Conservative Christians,* in which he argues that the Religious Right's combative tone is "a turnoff to reasonable people" and that it is "too limited in its emphasis and issues." After critiquing the Christian Right's motives, tactics, and strategies, he bluntly challenges its authority, writing that "we are looking for new leadership." Hunter was also a prominent supporter of the ECI, appearing in the national TV commercial that accompanied its announcement. His support was not coincidental, for expanding the ECI was consonant with his call for a broader agenda and more reasonable tone. And as *Right Wing, Wrong Bird* makes clear, he understood that his support for the ECI constituted a challenge to the Christian Right's power.[30]

Warren was less directly involved with advancing the ECI but was nevertheless among the most prominent and influential of its evangelical supporters. Like Hunter, he was a megachurch pastor and best-selling author. Indeed, having written a book called *The Purpose Driven Life,* which had sold twenty million copies by 2006, he was a rising evangelical star. In 2005, *Time* magazine had named him one of the "25 Most Influential Evangelicals in America," citing a poll of six hundred pastors that had listed him as second only to Billy Graham when it came to his ability to influence church affairs.[31] Warren's vision for Christian social engagement, like Hunter's, differed starkly from that advocated by the Christian Right. "I'm coming from the fact that Jesus said, 'Love your neighbor as yourself,'" Warren said at a 2005 news conference. "So what motivates me is not politics."[32] Indeed, rather than representing the Christian Right, Warren often challenged it, stating in an interview several months after the ECI launched, for example, that "there are a large group of people who are tired of the partisanship, the backbiting, the rudeness, the polarization in our country."[33] Warren's chief passion when he signed the ECI was combating poverty and AIDS, but his support for the ECI was of a piece with the alternative model of Christian social engagement that he promoted. Reflecting his prominence and the challenge it presented, news accounts of the ECI nearly always mentioned his participation.

Another movement that had begun to offer an alternative to the Christian Right's model for public engagement around this time is known as emerging evangelicalism.[34] Though amorphous, emerging evangelicalism began as a reaction against the religious culture of conservative evangelicals. In direct opposition to the Christian Right's model, emerging evangelicals reject "any definitive political allegiance or agenda."[35] That a leading figure in emerging evangelicalism, Brian McLaren, signed the ECI indicates another connection between evangelicals pushing for a broader agenda unbeholden to the Christian Right and the push for action on climate change. McLaren himself implicitly challenged the Christian Right's model, asking rhetorically in his 2007 book *Everything Must Change* why its preferred issues continued to claim evangelicals' attention:

> Why are so many religious people arguing about the origin of species but so few are concerned about the extinction of species? . . . If we religious people have exclusively seized on a couple of hot-button questions [evolution, abortion, and homosexuality], what other questions should we be thinking about that nobody's asking?[36]

Following a similar model, David Gushee's 2008 book *The Future of Faith in American Politics* explicitly argued that an evangelical center was emerging that was poised to challenge the Christian Right's dominance. According to Gushee, "this evangelical center (and to a lesser extent the left, but not the right) is winning the hearts and minds of younger evangelicals and thus represents the likely future of evangelicalism far more than the graying evangelical right." For Gushee, creation care was among the key issues that motivated the center.[37]

Statements like these, which accompanied the intense media coverage surrounding the ECI, contributed to the sense that leaders in the Christian Right no longer had an unassailable claim to represent the evangelical voice in politics. Citing Rick Warren's activism on poverty and AIDS, for example, a *USA Today* article observed approvingly in mid-2006 that the Christian Right's style of "unrelenting attacks" was being replaced by "another model . . . one that makes a compelling case that evangelical belief in the public square need not always divide us."[38] In 2007, a *New York Times* story similarly noted that the "fiery old guard who helped lead conservative Christians into the embrace of the Republican Party are aging and slowly receding from the scene." Warren (again) typified their replacement: a "new breed of evangelicals" who were "more likely to speak out about more liberal causes like AIDS, Darfur, poverty, and global warming than controversial social issues like abortion and same-sex marriage."[39] The boost that the ECI had given to the narrative of the Christian Right's weakening hold was unmistakable.

Equally unmistakable was the threat that this prospect posed to the Christian Right's power. As the 2008 election loomed, stories about "The Evangelical Crackup," "The Rise of the Evangelical Center," and "The New Evangelicals" who were set to challenge the Christian Right continued to appear.[40] So serious was the threat that by 2008 the Democratic presidential nominee, Barack Obama, was courting evangelical voters, a strategy Democrats had not been able to seriously consider for decades. Richard Cizik, vice president for governmental affairs at the NAE, even claimed in an early 2008 interview aired on National Public Radio that "forty percent of evangelicals are up for grabs. They could easily go for Obama." Obama was a "post–religious right kind of candidate," he added, twisting the knife.[41]

Demonstrating that the threat had registered, in 2008 Chuck Colson penned an opinion piece for the *Christian Post* dismissing the notion of an evangelical crack-up as "wishful thinking." Illustrating the significance of climate change in particular, he made a point of disputing the conclusion offered in a *New York Times* story that the issue of global warming was dividing evangelicals, calling it "a manufactured story"—even though Colson himself had been the lead signatory of a 2006 letter to the NAE (described below) requesting that it not take a position on climate change because the issue was, in fact, divisive.[42]

As Colson's evident alarm suggests, the specter of an evangelical electorate up for grabs was a critical factor driving leaders in the Christian Right to act to quash concern about climate change. If the Christian Right could not deliver votes, they could not count on Republicans to advance their agenda.[43] That the "new evangelicals" in the center still held many of the same positions regarding abortion and sexual morality as did their counterparts in the Christian Right did not matter; the new evangelicals' advocacy of a broader agenda would have directly threatened the Christian Right's ability to make progress on its issues. Cizik made the threat clear in his National Public Radio interview, identifying Pat Robertson and James Dobson as the "old guard" who were losing support, specifically because they were not "addressing issues like the environment" on which new evangelicals wanted leadership.

In this context, it is eminently understandable that the SBC's Ethics and Religious Liberty Commission threatened "to unleash the full scale of their arsenal" against the SBECI, whose "Southern Baptist Declaration on the Environment and Climate Change" not only contributed to the narrative of the greening of evangelicalism, but would have raised inconvenient questions, given its election-year timing, about the Christian Right's sway over the evangelical laity. It also would have suggested that the momentum unleashed by the ECI was unlikely to dissipate of its own accord. Countermeasures would be necessary.

The Secular Media and the Battle for Laypeople

A second way in which the ECI and the SEBCI threatened the interests of the Christian Right had to do with the intense media coverage of the ECI and, more broadly, of the new evangelicalism. Because of the lack of a centralized authority in evangelicalism, both evangelical environmentalists (and their new evangelical allies) and the Christian Right needed the support of lay evangelicals in order to credibly claim to represent the evangelical voice in the political arena. Therefore, the intense media interest that the ECI generated was a major problem for the Christian Right. In fact, evangelical environmentalists' ability to draw media attention was perhaps their singular advantage over the historically more powerful Christian Right.

Evangelical environmentalism's newsworthiness had been evident from its early days. Discussing the EEN's 1996 campaign against attempts to weaken the Endangered Species Act, one of the EEN's cofounders, Ron Sider, recalled that "we were a top story on the evening news. There were headline stories all across the country."[44] The same dynamic was evident in 2002 when the EEN first decided to broach the issue of climate change with its campaign "What Would Jesus Drive?" (WWJD). With an evangelical-friendly president in office, the Christian Right appeared to be on the verge of seeing some of its long-cherished policy goals implemented. But this triumph simultaneously ensured that the EEN's campaign generated intense media interest. To Sider's delight, the campaign generated five hundred TV segments and thousands of newspaper stories. "Nothing in our history has ever received so much national attention!" he cheered.[45] Jim Ball, who had been hired in 1999 to run the EEN, was reportedly "shocked by the instant and overwhelming attention that WWJD got." According to one interviewer, years later Ball was "still in awe" of the "media craze" it had ignited.[46]

The ECI both benefited from and heightened the media's interest in evangelical environmentalism. While many stories acknowledged that the issue was divisive, headlines often chose to highlight the activism itself, rather than the divisions: *Newsweek* lauded "God's Green Soldiers," while a *New York Times* editorial announced, "And on the Eighth Day, God Went Green."[47] The intense news coverage of the EEN's efforts indicated an unanticipated but convenient fact for evangelical environmentalists: evangelicals embracing the cause of global warming had the "man bites dog" quality that marks newsworthiness.[48] This gave them a distinct advantage over the Christian Right, whose opposition had the comparatively snooze-worthy "dog bites man" quality—a fact they would bemoan when they tried to get the secular news media to cover their efforts to oppose the ECI. Ironically, it was the Christian Right's very success in the larger realm of politics that gave evangelical environmentalists this advantage.

It is important to recognize that the media coverage of evangelical environmentalism was not merely inconvenient for leaders in the Christian Right. As noted above, the media coverage that evangelical environmentalists could leverage threatened the Christian Right's ability to pursue its agenda in the long run. Media coverage not only publicized what evangelical environmentalists were doing but also the arguments they were making. This helped them get their message out to the very people they needed to reach for their campaign to succeed, the evangelical laity. Of course, this was the same group whose allegiance the Christian Right needed in order to continue their alliance with Republicans.

That news stories about evangelical environmental activism could have a powerful effect on the evangelical laity was evident in my own field research. Allison, who headed the Georgia chapter of PCS (discussed in chapter 2), told me that her organization had focused on raising its visibility via media appearances of many kinds because "that's the biggest thing, that people know we exist." In raising the issue of creation care, in fact, she had learned that "there are lots of sort of hidden, closeted environmentalists all around us." Because of the social pressure to conform discussed in earlier chapters (also evident in Allison's choice of metaphors), such individuals tended to remain "closeted." She had discovered that when they learned about a form of environmentalism that was accessible to them as Christians, however, they were eager to get involved. Hence, visibility—perhaps most easily achieved via media appearances—was key to the PCS strategy of building the creation care movement. Allison went on to tell me of an eight-thousand-member evangelical church she had visited where a meeting with two members about starting a creation care group snowballed into fifty congregation members getting involved. (Interestingly, I had interviewed the pastor of the same church, and he had emphasized the importance of not imposing his environmental views because environmental concern was not "essential" to the faith, thus underlining his limited authority on non-consensus issues.) What PCS was able to do on a small scale in Georgia via local media appearances was only a shadow of what the ECI had the potential to do on a national scale. It was this prospect—millions of evangelical environmentalists coming out of the closet—that the Christian Right feared. And with good reason.

Climate Change and the Christian Right's Republican Coalition Partners

A third factor that pushed the Christian Right into the climate fight had to do with their coalition partners' opposition to action on climate change. Economic conservatives were deeply opposed to action because of the expense and increased government intervention it would entail. As the sociologists Robert Antonio and Robert Brulle explain, "anti-environmentalism has been, from the start, a keystone of neoliberal antiregulatory politics. But the perceived threats posed by climate

change discourse intensified this opposition, mobilizing energy companies and other related industries and broader free-market forces." Climate change quickly became the bête noire of conservatives, targeted
almost from the moment when it first aroused public interest in the late
1980s.[49]

Although neoliberalism gained traction in both parties, as time has
gone on, opposition to action on climate change has increasingly been
channeled through the Republican Party.[50] This made evangelicals, by
default, a core constituency for the climate opposition. In the early 2000s
evangelicals were relatively disengaged from the issue, but this did not
matter because they were not pushing for action. The advent of the ECI
in 2006 threatened to fundamentally alter this picture. The idea that
evangelicals might push for action on climate change was a direct threat
to the Christian Right's free-market allies in the Republican Party.[51] The
evangelical tradition's size and level of cultural influence would have
given an immense boost to the push for climate legislation, possibly the
critical push it needed to pass. Since evangelicals were a critical constituency, their defection on climate change would have been intolerable—
perhaps a deal breaker—for economic conservatives who opposed
action.[52] Hence, leaders in the Christian Right were compelled to get
involved in order to protect their coalition partners. As Bean and Teles
explain, "evangelicals are expected to do what they can to integrate their
belief system with that of their coalition partners, and to police unortho-
doxy within their own ranks when it threatens those partners."[53]

To gain some appreciation of evangelicals' clout, it is worth considering how thrilled environmentalists were about the prospect of evangelicals embracing "climate care," as evangelical climate activism has been
called.[54] "Lay evangelicals represent a traditionally disengaged—or
even opposed—audience that mainstream, secular environmentalism
has historically failed to reach," observed Katharine Wilkinson in 2012,
arguing that environmentalists should take seriously the prospect of
evangelical climate activism.[55] After the ECI launched, environmentalists' hopes soared ever higher. As Carl Pope, executive director of the
Sierra Club, wrote gleefully in the summer of 2007, "Being an environmentalist these days is like surfing the monster waves that periodically
pound the California coast. The power behind us is beyond our control.
The risks are enormous, as are the opportunities." While a columnist
for E-The Environmental Magazine celebrated the awakening of the
"sleeping giant" of the "Christian Right," the editor of the Ecologist
enthused optimistically that "it's hard to exaggerate the significance of

this shift. For many, the American religious right represents the last defense against political momentum to take climate change seriously, and its apparent U-turn will add hugely to the pressure already faced by the Bush administration."[56]

Of course, environmentalists were completely wrong in their assessment that the ECI represented the Christian Right's views on climate change. But they were correct that it amounted to waking a sleeping giant.

The Critical Role of the ECI and the SBECI

The Christian Right had multiple compelling reasons for wanting to discourage concern about climate change among lay evangelicals. But it is doubtful these reasons would have come into play if the ECI and the SBECI had not appeared. In promoting climate skepticism, leaders in the Christian Right would have to invest resources while also putting themselves at risk. The only reason they would have decided to engage, we can infer, is if the benefits outweighed the costs.

Prior to the appearance of the ECI, getting involved in the issue of climate change might have pleased some of the Christian Right's coalition partners in the Republican Party, but it would have been costly and risky to undertake. As for costs, promoting climate skepticism to the laity would be expensive, incurring costs in terms of man hours and air time. Directing supporters' attention to climate change would also distract them from the Christian Right's core issues. At the same time, delayed action on climate change offered no immediate or obvious benefits to their agenda.

As for the risks, climate change skepticism does not, on its face, have a clear connection to faith—at least not to the elements of faith that Christian Right leaders have historically tended to highlight. A sudden decision to address it would therefore appear to have suspiciously partisan origins. Certainly, the alliance with Republicans is no secret, but the basis of the alliance—opposition to big government—must make religious sense in order for it to be compelling to laypeople. And as we have seen, in the mid-2000s a number of voices were criticizing the Christian Right's overt partisanship, arguing that it had compromised the Christian witness. To suddenly tout an issue that was disconnected from the Christian Right's long-standing political agenda would be to put themselves at risk. Appearing too partisan would threaten their credibility with some lay evangelicals, thereby threatening the basis of their political power.

The appearance of the ECI and the SBECI would have dramatically altered this calculus. The costs and risks of mounting a campaign would still have been the same, but now there would have been great costs to *inaction* as well. The publicity the two initiatives attracted threatened to undermine the Christian Right's base of support among rank-and-file evangelicals; that they came out at all raised questions about the Christian Right's claim to leadership of the movement; and evangelicals' political clout would have worried neoliberal climate denialists in the Republican Party, who could be expected to exert pressure on their partners in the Christian Right. All of these factors would have tipped the balance in favor of action.

Two further pieces of evidence highlight the ECI's critical role in motivating the climate skepticism campaign. First, that leaders in the Christian Right would have preferred to avoid getting deeply involved in the issue if possible is evident from the fact that despite their alliance with anti-environmentalists in the Republican coalition since the 1980s, they had made little concerted effort to counter the activities of evangelical environmentalists, and until the ECI came out most of these were defensive, designed to counter what evangelical environmentalists were doing, rather than to actively advance their own position.[57] The EEN was founded in 1993, but it was not until 1999 that evangelical anti-environmentalists organized any kind of response, and even then, the organization they founded, the Cornwall Alliance, maintained a relatively low profile until the ECI appeared.[58] That this organization did not seek tax-exempt status or, apparently, hire full-time employees until 2009 underlines the significance of the SBECI.[59]

The second piece of evidence is a speech Senator James Inhofe gave at an event known as the Values Voter Summit, which was held in September 2006, about eight months after the ECI appeared. The summit drew an estimated seventeen hundred religious activists and included a panoply of influential figures in the Christian Right, including Jerry Falwell, Tony Perkins, and Donald Wildmon of the American Family Association.[60] It was an ideal way to reach religiously conservative activists, which Inhofe candidly admitted was his primary reason for presenting there. In a speech focused heavily on climate change, Inhofe referred obliquely to the surge of evangelical activism that had accompanied the ECI. Dismissing the possibility that it originated in genuine concern about the environment, he alleged that liberals had intentionally brought the issue to evangelical churches in order to sow discord. Unfortunately, he warned, some churches and associations were buying into the efforts.

He recommended three actions to halt what he apocalyptically characterized as an attempt to "destroy the core values that everyone in this room believes in": audience members should join the Cornwall Alliance's predecessor, the Interfaith Stewardship Alliance (described below), they should tell their churches the "truth" about global warming (that it was due to natural causes), and they should get the associations to which their churches belonged to do the same. "If you do this," he concluded to applause, "you'll be doing the Lord's work and He'll richly bless you for it."[61]

It is unclear how directly Inhofe's speech was linked to the campaign that eventually emerged, but his decision to present at the Values Voter Summit on this particular theme does demonstrate that the threat that the ECI posed had registered, and that opponents were looking for ways to get conservative, politicized evangelicals to mount a response. Notably, it also suggested a powerful theme for the campaign to come: opposing action on climate change was Godly activity—the Lord's work.

FACTORS ENABLING SUCCESS: ORGANIZATIONAL CAPACITY, MEDIA ACCESS, TRUSTED MESSENGERS, AND COMPELLING FRAMES

Leaders in the Christian Right were highly motivated to promote climate skepticism, but how were they able to outmatch evangelical environmentalists? In terms of being able to launch the campaign, the first step was to develop adequate *organizational capacity,* which came in the form of the Cornwall Alliance. The Cornwall Alliance was critical to coordinating their efforts in a way that maximized the impact of skeptics spread throughout the highly decentralized evangelical world.[62] The Cornwall Alliance also allowed evangelical skeptics to coordinate with other actors in the climate denial machine, efficiently sharing human and other resources, while also cross-pollinating between the secular and religious climate skeptic communities.

As organizational capacity developed, however, it became clear that it was not enough to simply build an equivalent to the EEN. Because of the media disadvantage that leaders in the Christian Right faced, they had to seek an alternative means of conveying their message to laypeople. That alternative turned out to be the *evangelical mass media.* Their control of this resource ultimately gave them a decisive advantage over evangelical environmentalists and largely helps explain why the latter were, in the end, outmatched.

In terms of making the campaign successful, in addition to having the secular climate denial machine's help reaching the politically conservative masses (which would include many evangelicals), leaders in the Christian Right had a unique advantage compared to secular climate communicators in that they were *trusted messengers* within the evangelical community (or certain sectors of it, anyway). This helped make their message convincing, even though they were not experts on the topic of climate change. Equally important was that they had at their disposal a number of *frames* that were highly resonant with their conservative evangelical audience. This enabled them to craft a message that the specific audience they needed to reach would find compelling. And their access to the evangelical mass media meant they could selectively target an audience that would be less troubled by evidence of partisanship to begin with, reducing the risk. In particular, the Christian talk radio format lent itself naturally to political topics like climate change.

Developing Organizational Capacity:
The Cornwall Alliance and Its Secular Allies

The key figure in the backlash against evangelical climate activists associated with the ECI is an Orthodox Presbyterian theologian by the name of E. Calvin Beisner. Having studied under the economist and noted environmental skeptic Julian Simon, Beisner blends a free-market approach with conservative theology to minimize the dangers posed by environmental problems.[63] Beisner's approach made him a natural ally of the Acton Institute, which in the late 1990s had been looking to create a conservative alternative to the religious environmental umbrella group to which the EEN belonged, the National Religious Partnership for the Environment (NRPE).[64] Staff members ultimately drew heavily on Beisner's work to draft the new organization's founding document, the "Cornwall Declaration on Environmental Stewardship."[65] With this declaration in hand, the Acton Institute then began soliciting Catholic, Jewish, and Protestant leaders to join the advisory board of the new organization, to be called the Interfaith Council on Environmental Stewardship (ICES). Formally announced in a national press conference on April 17, 2000, ICES has since been renamed twice, as the Interfaith Stewardship Alliance (ISA) in 2005 and as the Cornwall Alliance for the Stewardship of Creation in 2007 (usually shortened to "Cornwall Alliance"). ICES had two major goals: to "supply the media with a ready source for contrarian views on religion and the environment" and to mobilize grassroots opposition to groups like the EEN and its umbrella organization, the NRPE.[66] While both goals were critical, it was only during the second half of the decade that the Cornwall Alliance would, via its involvement with the Christian Right's climate change skepticism campaign, be successful with the second.

From the start, the Cornwall Alliance's supporters were among the most influential in the Christian Right, including not only those discussed in chapter 7, but other well-known names such as Donald Wildmon, Bill Bright (president of Campus Crusade for Christ International), Beverly LaHaye (founder of Concerned Women for America), and Richard Land (former president of the SBC's Ethics and Religious Liberty Commission).[67] As David Gushee observed in 2008, "it is almost as if the evangelical right has subcontracted its environmental portfolio to Beisner."[68] Despite its interfaith origins—which reflected its origins as a counterpart to the NRPE—as the battle front moved decisively to evangelicalism in the mid-2000s, the Cornwall Alliance increasingly directed its resources toward conservative Protestants. As of 2018, its advisory board listed no members with Jewish or Catholic affiliations.

Under Beisner's guidance, the Cornwall Alliance (typically with a staff of just two people, Beisner and an assistant) has become the primary public voice of evangelical climate skepticism, performing a range of functions, from organizing leaders and disseminating skepticism to the evangelical masses, to providing research materials and expert testimony to congress.[69] In this capacity it has worked closely with secular groups promoting climate skepticism. For example, Beisner is also on the advisory board of the Committee for a Constructive Tomorrow (CFACT), a conservative think tank that the sociologists Riley Dunlap and Aaron McCright credit with helping to "[meld] climate change denial into the faux populist rage of the Tea Partiers."[70] The Cornwall Alliance's document "A Call to Truth, Prudence and Protection of the Poor" was coauthored by Paul Driessen, who also serves as a senior policy analyst for CFACT and as an "expert" with the Heartland Institute, another conservative think tank that promotes climate denial.[71] Further illustrating the crossover between the secular and evangelical climate skeptic camps, Beisner participated in the Heartland Institute's "International Conference on Climate Change" in 2008, 2009, and 2010, and the Cornwall Alliance cosponsored the event in 2011.[72] Finally, the Cornwall Alliance has also worked with the Heritage Foundation, coproducing the anti-environmentalist documentary *Resisting the Green Dragon* with it in 2010.[73] In partnership with secular skeptics, the Cornwall Alliance has played a central role in coordinating skeptical voices within the evangelical community and representing those voices to the broader public.

While the Cornwall Alliance ultimately became an effective hub for coordinating the dissemination of climate skepticism to the laity—as we have seen, often by Beisner or his associates serving as guests on radio programs dedicated to the topic—this capacity developed only slowly. For the first few years after the ECI was launched, the Cornwall Alliance was more effective at the first of its twin goals, which was making sure observers acknowledged the existence of two forms of evangelical environmentalism, rather than simply describing the successes of the EEN. However, even in this regard it struggled, having more success in getting stories about the greening of evangelicalism to acknowledge the lack of consensus than in generating positive news coverage of its own initiatives.

In January 2006, for example, just before the ECI was launched, the ISA sent a letter to the NAE requesting that it "not adopt any official position on the issue of global climate change," and that NAE staff not give the impression that the NAE supported action.[74] This letter helped ensure that the initial news stories about the ECI registered it as a non-consensus issue. For example, the opening words of the *New York Times* story acknowledged that the ECI had been established "despite opposition from some of their colleagues."[75] The acknowledgment of opposition did not diminish the attention being lavished on the ECI, but it did constitute a small win. Rather than simply describing the growth of evangelical environmentalism, media accounts increasingly recognized the existence of two opposing camps.

As the intense media interest that the ECI would generate became evident, Christian Right leaders remained on the defensive. They continued to try to work behind the scenes, while increasingly recognizing the problem that media coverage of the ECI posed for them. In a March 2007 letter to the chairman of the board of the NAE, numerous conservative evangelical leaders, the majority of them associated with pro-family parachurch organizations, complained about media coverage of Cizik, who had become a forceful advocate for the ECI. "The liberal media has given wide coverage to Cizik's views and has characterized them as being representative of the NAE member organizations," they protested, implicitly acknowledging that media coverage of evangelical environmentalists threatened their interests. The letter also demonstrated a desire to reach lay followers, for Dobson (a signatory) posted it to CitizenLink.org, a website that Focus on the Family was running at the time to encourage political activism.[76] It was a first, halting step toward mobilizing the laity against action on climate change.

In a 2007 letter that went out to Focus on the Family supporters, Dobson again signaled his concern about how the secular news media was portraying the issue, begging readers "not to believe the propaganda, now being shouted by the media, that Americans don't care about the issue of abortion any more, and have 'moved on' to global warming and other issues. It is a distortion of major proportions."[77]

The need to bypass the secular media was becoming ever more apparent. Equally apparent was that the Cornwall Alliance's efforts would never receive the kind of publicity in the secular news media that the EEN's did. About six months after the ECI launched, the Cornwall Alliance hosted a news conference at the National Press Club in Washington, D.C., to announce "A Call to Truth, Prudence, and Protection of the Poor: An Evangelical Response to Global Warming," which was a "point by point refutation" of the ECI's "Call to Action."[78] It was the same venue where the ECI had been announced, but in contrast to the media blitz that accompanied its rival, the Cornwall Alliance's effort received just four media hits. Only one of them was in a major national paper, the *Washington Post,* and that story was just a 113-word blurb in the metro section.[79] Jerry Falwell cut to the heart of the problem in a blistering op-ed penned for WND.com in November of that year. "The problem is that when evangelicals jump on board with liberal groups that are advancing climate alarmism, the so-called major media is there to trumpet their action." Yet when the ISA responded, he complained, its efforts were "virtually ignored."[80]

The Cornwall Alliance made various efforts to reach the laity during this period, but it is unclear how successful such efforts were. The stated aim of the Cornwall Alliance's 2008 "We Get It!" campaign, for example, was to collect a million signatures in support of its "landmark" statement of climate skepticism (actually a 154-word statement repeating previous Cornwall declarations). The campaign's website claimed that it "has been endorsed by thousands of pastors, Christian leaders, policy-makers, and theologians, as well as a growing host of national and state organizations," yet eight years after the campaign had begun it listed only twelve signatories, all of which had preexisting relationships with the Cornwall Alliance.[81] In early 2017, an overhauled website appeared; it listed twenty-five names—still nowhere near the original goal.[82]

The Cornwall Alliance's *Resisting the Green Dragon* DVD, which was released on Earth Day 2010, was designed to be used as a tool for lay leaders. It included, for example, a "leader's guide" with "discussion questions and suggested practical applications suitable for churches, Sunday schools, families, classrooms, and small groups."[83] In the five years after its release the Cornwall Alliance sold or distributed just under an estimated one thousand copies.[84] However, none of my focus group members mentioned the organization, nor did those in a separate, Texas-based study.[85]

It is difficult to know how successful such efforts were at reaching laypeople, but one thing is clear. While they were under way, a less visible campaign was beginning to take shape in the evangelical mass media. In this campaign, unlike the ones just discussed, associates of the Cornwall Alliance produced content that reached millions.

Reaching the Laity: The Power of the Evangelical Mass Media

Broadcasting has long been key to conservative Protestants' success. In the 1930s, most major fundamentalist leaders had radio programs. By the 1940s, Charles Fuller's *Old-Fashioned Revival Hour* reached twenty million listeners a week, or about 15 percent of the U.S. population at the time. During this period preachers were able to purchase radio time, but that changed when an organization representing several mainline Protestant denominations convinced major distributors to eliminate paid religious programming. Fundamentalists realized that if they wanted to be able to continue to preach on the airwaves, they needed to organize. The result was the NAE, which in turn created the National Religious Broadcasters (NRB) to lobby on behalf of evangelical radio preachers.[86] Their efforts paved the way for the enormous success of evangelical mass media entrepreneurs in the present era.

Today some 90 percent of evangelicals consume evangelical media each month, and one in five *Americans*—not just one in five evangelicals—consumes this type of media daily.[87] Although Christian Smith's research was conducted in the mid-1990s, its findings regarding evangelical mass media consumption are impressive. The average number of hours his evangelical respondents spent listening to Christian radio programs per week was 6.9; the average number of hours spent watching Christian television programming per week was 2.6.[88] Another study, from the early 2000s, found that 63 percent of ministers surveyed listened regularly to religious radio, further demonstrating its influence.[89]

Despite the size and influence of the evangelical mass media, however, few people—myself included, before I began this research—realize what a vast institution this "electronic church" is. As of 2016, the evangelical media included a dozen TV networks carried nationwide via satellite and cable and more than 3,400 radio stations (one-fifth of the total in the United States), as well as the rapidly growing front of digital media. Communication scholar Mark Ward Sr. summarizes the tentacular reach of these media with the example of Salem Media Group (formerly Salem Communications), the top evangelical radio network in the United States. As he writes, Salem

> aggregates syndicated audio and video programming, streams it on the Internet, and generates 59 million monthly site visits, 24 million Facebook friends, 9 million email subscribers, and 6 million monthly mobile visitors. That outreach is *in addition* to its 100 major-market radio stations, 2,400 affiliate stations that carry Salem news, talk, and music programming in 300 markets; a Sirius XM satellite radio channel; and five Salem-produced conservative talk shows with a combined weekly audience of 11 million listeners.[90]

Conveniently for the leaders in the Christian Right who opposed action on climate change, at the time the ECI launched, most of them had already built impressive media empires. This symbiotic relationship between the Christian Right's major leaders and the evangelical mass media is no coincidence, for their media empires were in fact integral to their status as influential leaders, enabling them to build up a base of supporters. In an important contrast, no one in the evangelical environmentalists' camp had similarly unrestricted access to evangelical mass media outlets.[91] The result was that evangelical environmentalists could not speak directly to evangelicals on a massive scale in ways that would resonate specifically with that audience. Leaders in the Christian Right, meanwhile, had exactly that sort of platform at their disposal.

The popularity of the evangelical mass media's products has made these media "a decisive social institution for the evangelical movement."[92] Radio has been particularly powerful: via its many news/talk programs, Christian radio plays a crucial role in determining which issues will be salient to evangelical voters. As such, even in the era of digital media, radio broadcasting "still carries much of the weight in creating evangelical consensus on the movement's social and political agenda."[93] Secular observers often underestimate the evangelical mass media's influence because, beginning with the cable revolution of the 1990s, audiences fragmented and the evangelical mass media went "underground," their "fare distributed via religious channels consumed almost entirely by religious audiences."[94]

Nevertheless, I deem the evangelical media's existence critical to the Christian Right's ability to launch a campaign quickly and successfully, and to do so with little awareness on the part of non-evangelical observers.

Convincing the Laity: Trusted Messengers and Compelling Frames

With access to the evangelical mass media, leaders in the Christian Right were well positioned to reach lay evangelicals. Conveniently for them, they also had major advantages when it came to *persuading* that audience. Beyond climate change, in fact, evangelical leaders have managed to become trusted opinion leaders on a number of matters about which they have little expertise, from psychology to history to science. Randall Stephens and Karl Giberson attribute this to their unique ability to harness the anti-intellectual impulse in American society, which is rooted in the country's democratic impulse: if all votes count the same, then, implicitly, the common man on the street is as qualified to judge matters of policy as experts who have studied the issue extensively. This anti-intellectualism is especially strong within evangelicalism, where the idea that more education can be detrimental to faith has deep roots in eighteenth- and nineteenth-century revivals.[95]

Drawing on this potent anti-intellectual strand, leaders in the Christian Right have championed anti-elitism—and thereby implicitly championed their own superior guidance—through screeds against "secular elites" of all stripes (see chapter 4). This anti-elitist discourse was particularly useful when it came to climate change, for leaders could portray themselves as more trustworthy than "elite" scientists by virtue of their commoner status. Their ability to portray themselves as trustworthy was especially useful because with an issue like climate change, which is both technically complex and difficult to perceive at the individual level, some people will take the mental shortcut of assessing the messenger's trustworthiness, rather than trying to process complex or technical language themselves.[96] An additional advantage that Christian Right leaders had is that people are more likely to trust messengers from their own cultural group.[97] Having helped define the boundaries of the movement, leaders in the Christian Right were the ultimate insiders. This further heightened their credibility with their target audience.

With access to media and credible, trustworthy messengers, the Christian Right had two of the most important factors known to have shaped opinions about climate change on their side: media coverage and elite cues.[98] Their final advantage was that they were able to harness the power of what social scientists call "framing" to make their message persuasive. *Framing* refers to how a story is told about a particular issue. Which story line a communicator chooses is key, for it shapes how the audience will respond to it by "[setting] a specific train of thought in motion, communicating why an issue might be a problem, who or what might be responsible for it, and what must be done about it."[99]

In the case of climate change, leaders in the Christian Right adopted three primary frames. As my focus groups and other research have demonstrated, among the most important of these was the frame that Christians were under attack by secularists, a richly resonant narrative that evangelical leaders had employed for decades. The fusion of free-market ideology with the Christian faith, accomplished over the decades of the Christian Right–Republican alliance, was a second established frame they could usefully invoke. Finally, they could also appeal to evangelicals' long-standing suspicions of elites and scientists, and of paganism, pantheism, or New Age spirituality associated with the environmental movement. With just a few key words, climate communicators in the Christian Right could call to mind a whole host of negative associations that worked in their favor.

Green evangelicals were aware of the importance of framing as well, but their message entailed forging what for many lay evangelicals were new or unfamiliar connections between Christianity and environmental concern. Even if they had been able to use the evangelical mass media to the same extent their opponents did, time was on the side of the Christian Right, for it was quicker and simpler to play off of existing connections than to propose new ones.

Stephens and Giberson neatly summarize the mixture of factors that gave the Christian Right its persuasive advantage:

> Christian audiences, understandably bewildered by competing claims about the age of the earth, human origins, the nature of homosexuality [etc.] often hear two very different arguments. One argument comes from an unknown but well-credentialed scientist who works at a famous but very liberal university. . . . This case is made with no consideration of how this new information fits within the larger framework of Christian theology and whether the new facts challenge deeply held beliefs. The other argument comes from a fellow believer and is couched in specifically biblical terms; the more academic argument is critiqued as both uncertain and incompatible with Christian beliefs. The credentials and affiliations [of the Christian speaker] might not be as impressive, and the science/history/psychology might even be a bit thin. . . . But all the relevant cues indicate that the argument is trustworthy, largely because of who is making it.[100]

In the case of climate change, the very credentials, prestige, and expertise of climate scientists and their general inability to put their findings in Christian terms (factors that were either neutral or beneficial for non-evangelical audiences) worked against them for some evangelicals. By contrast, men like Chuck Colson, Ken Ham, or David Barton could use their credibility as righteous men of God to help persuade their listeners that accepting climate change conflicted with a biblical worldview and aided the very enemies that had been intently rooting Christianity out of public life over the past century. In so doing, they were following best practices in risk communication, affirming evangelicals' sense of identity—conveniently, an identity they themselves had helped forge.[101]

In sum, leaders in the Christian Right were strongly motivated to promote climate skepticism among the evangelical laity and had a number of tools at their disposal to do so successfully. Chapter 7 showed how vigorous and comprehensive that campaign was. But it was not just vigorous and comprehensive; it was also effective. Their efforts halted the ECI's momentum and, in so doing, helped thwart the passage of climate change legislation in 2009–10.[102] The boost their efforts gave to the secular climate denial machine's efforts also gave Republicans opposing action on climate change substantial cover through the 2016 election cycle, delaying attempts to address climate change during an ecologically critical window.[103] The threat of popular support for action on climate change averted, decisively so among evangelicals, the United States would announce its withdrawal from the Paris Agreement in 2017.

Epilogue

When I began the research for this book, the topic seemed to be a backwater—an intriguing one, but a backwater nonetheless. Apart from brief surges of interest in apocalypticism during the Cold War, in the early 1980s with the emergence of the Christian Right, and at the turn of the millennium, end-time beliefs have drawn the attention of only a small number of scholars. While imagining that I would unobtrusively investigate a corner of the evangelical world of interest primarily to environmentalists, my research led elsewhere—to the confluence of religion and politics, waters that have in recent years grown especially turbulent. A number of critical insights have emerged from this unexpected journey into one of the most scrutinized arenas of American public life.

In academic terms, the most important is the extent to which the environmental attitudes of the subset of evangelicals I examined were shaped not simply by doctrine, but by history, identity, and community. This finding is significant because the social world has too often been ignored in attempts to understand evangelicals' environmental attitudes. It is also significant because it complicates our understanding of how theology acts to shape environmental attitudes. Evangelicalism has long been presented as the quintessential example of how religious doctrine shapes environmental attitudes: following the logic of the Lynn White thesis, evangelicals are thought to be less environmentally concerned than other Christians because they grant greater authority to scripture, including its teachings about human dominion over the earth.[1] Yet my work suggests an alternative reading of this finding: it is not that certain scriptural passages have a universal effect, but that evangelicals' experience in American culture has encouraged particular practices of scriptural interpretation.[2] In the case of climate change, skepticism was not simply what happened when evangelicals considered the issue in light of scripture, but in part the result of certain evangelical leaders and pundits having an interest in portraying skepticism as the scriptural view.[3]

It is difficult to predict the long-term impacts of their efforts, but in the short term they paid off handsomely for the Christian Right's coalition partners, in the form of delayed action on climate change. Christian Right leaders have also benefited in terms of continued political leverage. It seems likely that in the long run, however, continued action of this kind could erode their credibility with the laity, ultimately undermining the basis of their power.

More generally, while the period of history this book covers was a discouraging one for climate activists, this insight into the shaping of evangelicals' environmental attitudes is ultimately a source of optimism, underlining that—as the efforts of some younger evangelicals indicate— a vibrant evangelical environmentalism is possible.[4] Since environmental attitudes are deeply rooted in culture—that is, in perceptions, values, identities, and social relationships—it does not seem likely that the attitudes of traditionalist evangelicals will change quickly or easily. But centrist evangelicals may be more open to persuasion.

A second corrective is related to how political conservatism shapes evangelicals' attitudes toward climate change. In the popular press it is commonly asserted that, to the extent that evangelicals are climate skeptics, it is primarily because they are politically conservative. It is true that quantitative analyses have shown that political orientation has a much stronger effect on climate change attitudes than factors such as demographics (including religion), cultural orientation, or personal experience.[5] However, this formulation relies on a problematic view of politics as separate from religion. In reality, a close connection between evangelical religiosity and conservative politics has developed in the United States since the late 1970s, with the result that today many white evangelicals are politically conservative because of their religious beliefs. To consider political orientation as a separate, nonreligious driver of climate change attitudes for this group is to underestimate the true impact of their religiosity.[6]

My perspective from the field underscores this point. My informants consistently described their political views as having religious origins.

A teacher in her twenties who participated in one of my focus groups explained, for example, that "my political views are mainly biblical. I don't go for just Democrat or Republican, but if the views are pleasing to God." Another teacher, in her thirties, told me that "my politics are in relation to the Bible." More specifically, she liked "some aspects of the Democratic party" but voted Republican because as a Christian she was pro-life and opposed gay rights.

A particularly clear example of how climate skepticism was rooted not only in politics but in the combination of religion and politics can be seen in my finding that hot millennialists typically accepted that the climate was changing, while cool millennialists were predominantly skeptics. The narrative traced throughout the book suggests why this is the case. As discussed in chapter 3, hot millennialists tended to be less politically engaged, withdrawing from a world they saw as wicked beyond redemption. As a result, they were less likely to encounter the skeptical discourses that the conservative media disseminated (chapter 4) and less subject to the social pressure to conform that went along with them (chapter 5).[7] This also meant they were less likely to have been exposed to the skeptical messages that Christian Right leaders disseminated (chapter 7). All these factors made it a straightforward matter for them to accept that the climate was changing. On the other hand, cool millennialists were keenly aware that climate change was a liberal issue, were typically surrounded by others who felt the same way, and, in many cases, were suspicious that accepting anthropogenic climate change was incompatible with a biblical worldview. As a result, they largely rejected it.

In the case of climate change, politics clearly plays an important, direct role in shaping attitudes—this much is clear from research on Americans at large.[8] But as I have shown, among traditionalist evangelicals, politics shapes attitudes toward climate change both directly *and* indirectly. Unlike secular conservatives, many traditionalist evangelicals receive a double dose of climate skepticism via politically conservative elites and media *and* from religiously conservative elites and media. Coming full circle, many of those religious elites explicitly present politically conservative views as the natural expression of a biblical worldview.[9] The tendency to attribute white evangelicals' climate skepticism primarily to politics thus underplays the many ways in which their politics and religiosity are intertwined.[10]

My research also sheds new light on the quantitative studies that have found a link between end-time belief and environmental attitudes. As mentioned in chapter 1, in their 1995 study, Guth and his colleagues found an especially strong link between end-time beliefs and environmental apathy among the politically mobilized.[11] My research suggests that this may be due, in part, to the nature of conservative evangelicals' political involvement: becoming politically active would likely have exposed them to anti-environmentalist coalition partners in the Republican Party. Others may have opposed environmental regulation for ideological reasons that dovetailed with their faith, as in the case of James Watt. Because they were motivated to enter politics in order to defend traditional Christian faith, these Christians would have been especially likely to hold orthodox eschatological views, which explains the particularly strong association between end-time belief and reduced environmental concern among them.

Supporting the idea that ideological drift from evangelicals' New Right coalition partners played a role, in a separate analysis of one of the data sets that contributed to the 1995 study, Guth and his colleagues found that religious activists associated with the Christian Right were less willing than those in the Christian Left or Center to support environmental policies that might cause inflation or cost workers their jobs.[12] The authors also found that only 3 percent of respondents belonging to the Christian Right identified the environment as the "most important [societal] problem," compared to 47 percent of individuals in the Christian Left and 27 percent in the Christian Center, which seems to illustrate the powerful impact of political mobilization on issue attention. This is not to claim that end-time beliefs have no impact on environmental attitudes, of course, but rather that some of the variance that has been attributed to otherworldly indifference may, in fact, be attributable to this-worldly engagement.

As for Barker and Bearce's study from 2012, my fieldwork suggests an alternative interpretation of its findings as well. The majority of my informants were theologically conservative and therefore would likely have answered yes to the very broad survey question used to measure end-time beliefs, "Do you believe in the *Second Coming* of Jesus Christ—that is, that Jesus will return to earth someday?"[13] Only a small number of them would have interpreted this to mean that preserving the earth for future generations was futile, but this did not mean that they would have given the pro-environmental answer to the study's question about climate change, which asked respondents the degree to which they agreed that "global warming is a problem that requires immediate government action, in order to prevent environmental devastation and catastrophic loss of life for future generations."[14] Beyond embattled evangelicals' tendency to be skeptical about human-caused climate change to begin with, two aspects of this question would have been particularly likely to encourage a negative response from this group: the reference to the need for *government* action and the alarmist (as it would be interpreted) reference to "catastrophic loss of life." Like the earlier study by Guth and his colleagues, Barker and Bearce's findings about the impact of end-time beliefs may also implicate the embattled mentality to some degree.

On a different note, throughout my research I was struck by the parallel dynamics that I saw in evangelical churches and in the environmentalist circles of my acquaintance. I am thinking, in particular, of the ways in which evangelicals framed environmentalists as an out-group, which enhanced their sense of solidarity, mapping their identity so as to exclude environmentalists and thereby present themselves in a positive light. In my experience, some environmentalists also view evangelicals as an out-group—one whose naive faith had allowed their tradition to be co-opted for political purposes. If evangelicals contrasted themselves with environmentalists by highlighting their command of common sense, environmentalists contrasted themselves with evangelicals by highlighting their commitment to science and reason and their superior grounding in reality. This social distance is a powerful but generally unacknowledged barrier to building a movement that crosses political lines.

Curious whether my jarring experience in moving between evangelical and environmental worlds was part of a broader pattern, I asked Allison from PCS, the organization that worked to bridge religious and environmentalist communities in Georgia, for her thoughts. She laughed good-naturedly and told me that in her experience, "the environmental community is just as judgmental as those types of Christians who say, 'How could you be a Christian and an environmentalist?'" In fact, she told me that she had found it just as difficult to convince her environmentalist partners in Georgia to trust Christians as it was to convince Christians to get involved with environmental issues. As she put it, with a sense of irony that was clearly much needed in her position, "I've had to spend the past four years of having this job convincing *both* sides that we're not the devil."

In the mid-2000s, environmentalists were elated that they might be able to get evangelicals' support in their push for legislative action on climate change. But I suspect they were not fully cognizant of the socially powerful boundaries between the two communities. In retrospect, the fact that the celebratory "greening of evangelicalism" narrative emerged at almost exactly the same moment as Bill Moyers's widely circulated "Welcome to Doomsday" speech about the environmental threat posed by apocalyptic evangelicals underlines environmentalists' profound and largely unexplored ambivalence toward evangelicals. While the evangelical mass media have done much to distort evangelicals' understanding of environmental issues and thereby to derail the evangelical environmental movement, the *mutual* disregard between the two groups has surely been a barrier as well.

If I am correct that end-time beliefs play a somewhat peripheral role in driving apathy about climate change, then depicting end-time believers as villains—antagonists who must be attacked and defeated—is problematic. To adopt such a stance is not only to accept but to deepen the societal divisions that conservatives carved out with the rise of the New Right and the Christian Right. And this divisiveness is advantageous to environmentalists' opponents. Perpetuating the us-versus-them dynamic makes it harder, rather than easier, for moderate, pro-environmental evangelicals to find a home in environmental circles. And it makes it harder, rather than easier, for non-evangelical environmentalists to welcome them. It convinces both sides that the other side is the devil.

As for practical lessons, this book provides many reasons to think that the prospects are dim, in the near term, for American evangelicals to add significant momentum to the climate fight. In the broader challenge of caring for the environment, however, there would seem to be room for alliance, especially on issues that affect human health. Here the Christian Right may even have a lesson to offer environmentalists.

Survey research has shown that there is a widespread desire for a clean environment, even among those who are politically conservative.[15] Notably, even my religiously conservative informants in Georgia shared this desire. Despite theoretically having a large base, however, environmentalists have been largely unable to join forces with politically and/or religiously conservative environmentalists—or "conservationists," as some might prefer to be called.

This situation calls to mind the situation of evangelicals before the rise of the Christian Right. As late as the 1960s and early '70s, large numbers of evangelicals shared similar beliefs, yet they did not think of themselves as one community that could and should articulate a unified political voice. The story of their mobilization is complicated, but a consistent theme is the savvy use of media. The NAE was originally formed to protect the rights of evangelical radio preachers, for example.[16] Later, in the early and mid-twentieth century, network radio played a key role in creating "a broadly coherent [evangelical] subculture."[17] Today, Christian radio continues to bind the evangelical community while also helping to set white traditionalist evangelicals' social and political agenda.[18]

Media were also key to the New Right's success. As Richard Viguerie observed in 1981, in the 1970s liberals had "effective control of the mass media" and, as a result, were able to "present the positive side of liberal causes, liberal issues, liberal personalities and, for the most part, ignore conservative causes, conservative issues, and conservative personalities or present them in an unfavorable manner." Direct mail—unsolicited advertisements mailed directly to consumers—was his solution to this problem, one of the four pillars of the New Right's success.[19] It was, as he put it, the "one method of mass commercial communication that the liberals do not control." He assessed its utility in no uncertain terms: "Without the mail, most conservative activity would wither and die."[20]

In the absence of a new media strategy, environmentalists may face the same fate. Environmentalists have a compelling message, but the movement has relied heavily on the secular mainstream media to cover its efforts and convey its messages. As the news media have become increasingly balkanized, with people self-selecting news that reinforces preexisting views, this strategy has lost its effectiveness.[21] My informants' inaccurate understanding of environmentalism and climate change is a clear sign of this.[22] The Christian Right's ability to outmaneuver evangelical environmentalists with their preferred message on climate change is another.

Those who want a clean, safe environment are in the moral majority. Yet, like evangelicals before Jerry Falwell popularized that phrase, they don't know it.[23] A long-term strategy for environmental resurgence may need to include finding new ways to activate this silent majority of people—including many evangelicals—who do not want to drink polluted water, breathe polluted air, or put the health of their children at risk because of lax environmental enforcement.

Methods

Focus groups are a research method in which a group of five to eight participants discuss a topic among themselves under the direction of a moderator. I selected focus groups for the first phase of research because they are well suited to assessing people's perceptions and feelings about a particular topic, uncovering differences in perspectives between groups or categories of people, and revealing factors that influence opinions, behavior, and motivation.[1] Importantly, in group settings participants can react to and challenge each other, providing a more realistic account of how groups construct meaning in their everyday lives.[2]

To identify evangelicals, researchers use three basic strategies: self-identification, denominational affiliation, or beliefs. My desire to examine environmental attitudes in the social setting of the church dictated a denominational strategy. To ensure that my informants had been shaped by the same congregational environment, I further chose to assemble groups composed of members from the same church. Working with people who knew each other was also beneficial because it meant that my focus group participants were generally comfortable discussing the topic of climate change (since propriety generally proscribes discussing controversial topics with strangers, groups whose members did not know each other would have been less forthcoming).[3]

I conducted focus groups between July 2011 and February 2012. I sought a diversity of denominations, as well as diversity in terms of location (rural, suburban, or urban settings) and church size. To create a sampling frame, I used local phone books to identify all the churches in my region, which included a town of about thirty thousand people ("Mayfield"), several small towns located in rural areas, and a city of over fifty thousand ("Dixon"). I narrowed down this list by excluding mainline, black Protestant, and Roman Catholic churches, which left white evangelical Protestants.[4] I then further narrowed down the list by identifying churches that were theologically conservative, a designation I determined by reviewing church or denomination websites for words and phrases like *inerrant, infallible,* and *inspired word of* God. I did this in order to hold theological interpretation relatively constant, so that it would not need to be considered as an explanatory factor for changing levels of environmental concern.

I next identified premillennialist and amillennialist churches on the basis of statements of faith found on church or denomination websites. I later confirmed this designation through interviews. I then began calling pastors to request interviews. My requests were sometimes ignored (even after repeated attempts) but were denied in only three cases. Ultimately, I conducted fifteen interviews with pastors and was able to arrange focus groups at just under half of the corresponding churches (some pastors I interviewed were unwilling to let me hold a focus group at their church). Table A.1 shows the denominations of the pastors I interviewed (multiple entries per denomination indicate that I interviewed more than one pastor from that denomination).

I relied on the pastors to help me identify participants, which they typically did by announcing the discussion in a church bulletin or during a service or Bible study. I asked the pastors to describe the focus group as a group discussion about the relationship between the Christian faith and values, rather than mentioning the environment or climate change, so as not to attract participants with particularly strong opinions about the topic. While not biased in terms of my informants' environmental attitudes, my sample is biased toward frequent church attenders, since they were more likely to be present on days when focus groups were announced. This is one reason why I consider my informants to be representative of the "traditionalist" subset of evangelicals—who tend to be frequent church attenders, while also sharing a number of other characteristics as discussed in the introduction—rather than of evangelicals in Georgia more broadly speaking.

To maintain rigor, I sought negative and qualifying evidence throughout the research process. Since my initial hypothesis was that end-time belief reduced concern about the environment, I sought negative evidence by examining the role of end-time beliefs in a number of denominational settings (i.e., not just among fundamentalist dispensational premillennialists). I also sought qualifying evidence by asking questions about the role of four other factors besides end-time beliefs: (1) political orientation; (2) attitudes toward science; (3) other this-worldly priorities, such as opposing abortion and gay marriage; and (4) tensions with environmentalists. A professional focus group facilitator helped me design a questioning route that would elicit discussion on all four topics. Prior to the first focus group, I also used cognitive debriefing in order to refine my questions.[5]

Focus groups lasted between sixty and ninety minutes and occurred on church premises after Sunday or Wednesday night services. I served as moderator, while an assistant took notes. At the end of each session, my assistant and I debriefed

TABLE A.1   SUMMARY OF INTERVIEWS

Denomination or organization	Pseudonym[a]	Date of interview
	Phase 1: Evangelicals	
Baptist		
Independent	Roger	July 19, 2011
	[Jim]	January 17, 2012
Southern Baptist Convention	Richard	January 11, 2012
	[John]	October 13, 2011
Missionary Baptist (American Baptist Association)	Bob	July 19, 2011
Christian Church	Joseph	October 18, 2011
	[Mike]	January 17, 2012
Churches of Christ	Joshua	July 5, 2011
Lutheran, Missouri Synod	[Luke]	January 18, 2012
Nondenominational	Phillip	January 4, 2012
Pentecostal		
Assemblies of God	William	October 5, 2011
Church of God	Ben	December 13, 2011
Nondenominational	[Aaron]	January 4, 2012
Presbyterian Church in America	Chris	October 26, 2011
Seventh-day Adventist	Matthew	January 25, 2012
	Phase 1: Environmentalists	
National Environmental Coalition (pseudonym)	Jim	April 6, 2012
	Cathy	April 23, 2012
	Kara	July 18, 2012
	[Sonya]	July 23, 2012
Clean Water Coalition (pseudonym)	Deborah	July 18, 2012
	[Linda]	August 14, 2012
Partners for Christian Stewardship (pseudonym)	Allison	March 27, 2012
	Phase 2: Southern Baptists	
Southern Baptist Convention	Jim	July 24, 2012
	[Connor]	July 25, 2012
	Matthew	July 25, 2012
	John	July 26, 2012
	Fred	July 27, 2012
	[Daniel]	July 30, 2012
	Mike	July 30, 2012
	Paul	July 31, 2012
	Greg	July 31, 2012
	Aaron	August 2, 2012
	[Christian]	August 2, 2012
	Jonathan Merritt	August 10, 2012

[a] Brackets indicate that the individual is not quoted in the book. All individuals except Jonathan Merritt are identified via pseudonyms.

to check our perceptions of what we had heard and of differences between groups. To increase confidence in the validity of the results, in the end of each discussion I read a summary of what I had heard back to the participants, adopting participants' language as much as possible, and asked them to clarify if anything sounded incorrect. I further asked participants to write down anything they wanted to say that did not come up or that they did not feel comfortable bringing up in the group discussion, in order to learn whether alternative views were being withheld. Finally, they completed a fifteen-item questionnaire, which asked their age, race, sex, education, frequency of church attendance, Bible views (using an item adapted from the General Social Survey tapping biblical literalism), the region of the country where they had spent most of their lives, their occupation, whether they were in the military, how they would describe their political views (open-ended), their self-rating of political ideology (with seven options ranging from extremely liberal to extremely conservative), what political party they tended to vote for, and their total family income.

I transcribed and coded the interview and focus group tapes as the focus groups were being conducted, following techniques developed by Strauss and Corbin.[6] This involved identifying common concepts throughout the transcripts, which are then grouped into categories for the purposes of theory building. Collecting and analyzing data simultaneously allowed me to test new insights in later focus groups, so that I could test new hypotheses as they emerged.

For the interviews with environmentalists, I made a list of all the environmental organizations that worked in the local region (including Mayfield, Dixon, and surrounding cities). I found just three: one national (the NEC) and two local (the CWC and another not discussed in this book).[7] As table A.1 shows, I conducted semi-structured interviews with leaders and officers in each of these groups to learn about the group's work and its involvement with and perceptions of local religious groups. I also interviewed the executive director of PCS, which was based in another part of the state, in order to better understand whether what I observed in my region was reflective of broader patterns in the state. Although I interviewed only a small number of environmentalists, they represent the voices of those who were most involved in organized environmental activism in the region, insofar as I could determine.

PHASE 2: CLIMATE SKEPTICS IN THE SOUTHERN BAPTIST CONVENTION

Signatories of "A Southern Baptist Declaration on the Environment and Climate Change" were identified from the list posted at www.baptistcreationcare .org as of January 12, 2012 (this website is no longer active). From a list of over four hundred signatories, I created a short list of signatories who were pastors located in Georgia or (to create a slightly larger pool) in adjacent regions of northern Florida and South Carolina. I chose to contact pastors rather than laypeople because I could typically find contact information for pastors by searching online for their churches. Only names and locations were available for lay signatories, which was generally not enough information to be able to contact these individuals. This process resulted in a list of twenty individuals, twelve of whom I was able to reach and interview by telephone between July 24 and August 10, 2012, including Jonathan Merritt. In these interviews, which were semi-structured in order to allow me to probe when necessary, I asked about my informants' decision to sign on to the Declaration in order to learn what led these pastors in particular to take an active stance. The realization that most of my signatories were climate skeptics prompted me to use the tools of cultural analysis described in chapter 6.

PHASE 3: ANALYZING EVANGELICAL MASS MEDIA COMMUNICATIONS

I analyzed the Christian Right's climate skepticism campaign primarily using digital archives. Extensive archives were available on the websites of CBN, *Break-Point, Answers with Ken Ham, Washington Watch* and *Live with Tony Perkins* (both at www.frc.org), *WallBuilders Live!*, Movieguide.org, and *Knowing the Truth*. I identified content related to climate change on these websites using search functions and the keywords *climate change* and *global warming*, with duplicates removed. This strategy likely underestimated total mentions of climate change, which also occurred with some frequency in brief or peripheral context (as indicated by limited purposive sampling). *BreakPoint*'s search function for 2011 and 2012 did not function properly as of 2018, leading me to limit my analysis to 2009–10 (although not included in the formal analysis, I did find two stories in 2013, one in 2014, and seven in 2015). For all other programs, I analyzed content for as many years as were available and complete between 2006 and 2015.

For *Generations Radio,* my first analysis, which was conducted in 2016, used the search function on the website Generationswithvision.com, but that website was no longer active by 2018. I was able to redo the analysis using *Generations Radio*'s posts on SermonAudio.com, using the methodology just described.

Because digital archives were lacking, I assessed Falwell's *Old-Time Gospel Hour* television and radio programs using news stories published about his sermons on climate change (see chapter 7). I assessed Kennedy's *Truths That Transform,* Barton's *WallBuilders Live!* (2008 only), and Wheaton's *The Christian Worldview* using archives available in the Internet Archive Wayback Machine (https://archive.org). Records are incomplete for Falwell's and Kennedy's programs, but those that are available make it clear that they both promoted climate skepticism.

For *Answers with Ken Ham* (archives for which are available at https://answersingenesis.org/media/audio/answers-with-ken-ham-volumes/), the story counts include not only the features about climate change that I identified using keywords but also stories to which advertisements of materials promoting climate skepticism were appended. The announcer for these advertisements typically provided a conceptual link between the biblical worldview presented in the feature and the materials promoting climate skepticism to which he was directing listeners, justifying their inclusion in the story count.

Once climate change stories were identified, the author or a research assistant divided them into three categories. *Skeptical* stories either contained misinformation about climate change (i.e., that there is no scientific consensus, that temperature increases correspond to sunspots, or other arguments listed on SkepticalScience.com's "Arguments" page) or focused on the negative results of addressing climate change.[8] *Accepting* stories provided information that reflected the scientific consensus or positive coverage of climate change advocates, proposals, or activism. Articles that had no clear stance were categorized as *neutral* or *unclear*. In limited cases where audio files were not available (such as for some archives accessed on the Internet Archive WayBack Machine), I categorized stories based on the story title and the identity of the guest speaker, with stories being assigned to the skeptical category only if the title clearly identified the topic as climate change or global warming and the speaker could be verified as having consistently promoted climate skepticism in other venues.

A Southern Baptist Declaration on the Environment and Climate Change

PREAMBLE

Southern Baptists have always been a confessional people, giving testimony to our beliefs, which are based upon the doctrines found in God's inerrant word— the Holy Bible. As the dawning of new ages has produced substantial challenges requiring a special word, Southern Baptist churches, associations and general bodies have often found it necessary to make declarations in order to define, express and defend beliefs. Though we do not regard this as a complete declaration on these issues, we believe this initiative finds itself consistent with our most cherished distinctives [sic] and rooted in historical precedent.

The preamble to the *Baptist Faith and Message 2000* (BFM 2000) declares: "Each generation of Christians bears the responsibility of guarding the treasury of truth that has been entrusted to us [2 Timothy 1:14]. Facing a new century, Southern Baptists must meet the demands and duties of the present hour. New challenges to faith appear in every age."

We recognize that God's great blessings on our denomination bestow upon us a great responsibility to offer a biblically-based, moral witness that can help shape individual behavior, private sector behavior and public policy. Conversations like this one demand our voice in order to fulfill our calling to engage the culture as a relevant body of believers. Southern Baptists have always championed faith's challenges, and we now perpetuate our heritage through this initiative.

We are proud of our deep and lasting commitments to moral issues like the sanctity of human life and biblical definitions of marriage. We will never compromise our convictions nor attenuate our advocacy on these matters, which constitute the most pressing moral issues of our day. However, we are not a single-issue body. We also offer moral witness in other venues and on many issues. We seek to be true to our calling as Christian leaders, but above all, faithful to Jesus Christ our Lord. Therefore, our attention goes to whatever issues our faith requires us to address.

We have recently engaged in study, reflection and prayer related to the challenges presented by environmental and climate change issues. These things have not always been treated with pressing concern as major issues. Indeed, some of us have required considerable convincing before becoming persuaded that these are real problems that deserve our attention. But now we have seen and heard enough to be persuaded that these issues are among the current era's challenges that require a unified moral voice.

We believe our current denominational engagement with these issues have [sic] often been too timid, failing to produce a unified moral voice. Our cautious response to these issues in the face of mounting evidence may be seen by the world as uncaring, reckless and ill-informed. We can do better. To abandon these issues to the secular world is to shirk from our responsibility to be salt and light. The time for timidity regarding God's creation is no more.

Therefore, we offer these four statements for consideration, beginning with our fellow Southern Baptists, and urge all to follow by taking appropriate actions. May we find ourselves united as we contend for the faith that was delivered to the saints once for all. *Laus Deo!*

STATEMENT 1

Humans Must Care for Creation and Take Responsibility for Our Contributions to Environmental Degradation.

There is undeniable evidence that the earth—wildlife, water, land and air—can be damaged by human activity, and that people suffer as a result. When this happens, it is especially egregious because creation serves as revelation of God's presence, majesty and provision. Though not every person will physically hear God's revelation found in Scripture, all people have access to God's cosmic revelation: the heavens, the waters, natural order, the beauty of nature (Psalm 19; Romans 1). We believe that human activity is mixed in its impact on creation—sometimes productive and caring, but often reckless, preventable and sinful.

God's command to tend and keep the earth (Genesis 2) did not pass away with the fall of man; we are still responsible. Lack of concern and failure to act prudently on the part of Christ-followers reflects poorly to the rest of the world. Therefore, we humbly take responsibility for the damage that we have done to God's cosmic revelation and pledge to take an unwavering stand to preserve and protect the creation over which we have been given responsibility by Almighty God Himself.

STATEMENT 2

It Is Prudent to Address Global Climate Change.

We recognize that we do not have any special revelation to guide us about whether global warming is occurring and, if it is occurring, whether people are causing it. We are looking at the same evidence unfolding over time that other people are seeing.

We recognize that we do not have special training as scientists to allow us to assess the validity of climate science. We understand that all human enterprises are fraught with pride, bias, ignorance and uncertainty.

We recognize that if consensus means unanimity, there is not a consensus regarding the anthropogenic nature of climate change or the severity of the problem. There is general agreement among those engaged with this issue in the scientific community. A minority of sincere and respected scientists offer alternate causes for global climate change other than deforestation and the burning of fossil fuels.

We recognize that Christians are not united around either the scientific explanations for global warming or policies designed to slow it down. Unlike abortion and respect for the biblical definition of marriage, this is an issue where Christians may find themselves in justified disagreement about both the problem and its solutions.

Yet, even in the absence of perfect knowledge or unanimity, we have to make informed decisions about the future. This will mean we have to take a position of prudence based partly on science that is inevitably changing. We do not believe unanimity is necessary for prudent action. We can make wise decisions even in the absence of infallible evidence.

Though the claims of science are neither infallible nor unanimous, they are substantial and cannot be dismissed out of hand on either scientific or theological grounds. Therefore, in the face of intense concern and guided by the biblical principle of creation stewardship, we resolve to engage this issue without any further lingering over the basic reality of the problem or our responsibility to address it. Humans must be proactive and take responsibility for our contributions to climate change—however great or small.

STATEMENT 3
*Christian Moral Convictions and Our Southern Baptist
Doctrines Demand Our Environmental Stewardship.*

While we cannot here review the full range of relevant Christian convictions and Baptist doctrines related to care of the creation, we emphasize the following points:

- We must care about environmental and climate issues because of our love for God—"the Creator, Redeemer, Preserver and Ruler of the Universe" (BFM 2000)—through whom and for whom the creation was made. This is not our world, it is God's. Therefore, any damage we do to this world is an offense against God Himself (Gen. 1; Ps. 24; Col. 1:16). We share God's concern for the abuse of His creation.

- We must care about environmental issues because of our commitment to God's Holy and inerrant Word, which is "the supreme standard by which all human conduct, creeds and religious opinions should be tried" (BFM 2000). Within these Scriptures we are reminded that when God made mankind, He commissioned us to exercise stewardship over the earth and its creatures (Gen. 1:26–28). **Therefore, our motivation for facing failures to exercise proper stewardship is not primarily political, social or economic—it is primarily biblical.**

- We must care about environmental and climate issues because we are called to love our neighbors, to do unto others as we would have them do unto us and to protect and care for the "least of these" (Mt. 22:34–40; Mt. 7:12; Mt. 25:31–46). The consequences of these problems will most likely hit the poor the hardest, in part because those areas likely to be significantly affected are in the world's poorest regions. Poor nations and individuals have fewer resources available to cope with major challenges and threats. Therefore, "we should work to provide for the orphaned, the needy . . . [and] the helpless" (BFM 2000) through proper stewardship.

Love of God, love of neighbor and Scripture's stewardship demands provide enough reason for Southern Baptists and Christians everywhere to respond to these problems with moral passion and concrete action.

STATEMENT 4
It Is Time for Individuals, Churches, Communities and Governments to Act.

We affirm that "every Christian should seek to bring industry, government and society as a whole under the sway of the principles of righteousness, truth and brotherly love" (BFM 2000).

We realize that we cannot support some environmental issues as we offer a distinctively Christian voice in these arenas. For instance, we realize that what some call population control leads to evils like abortion. We now call on these environmentalists to reject these evils and accept the sanctity of every human person, both born and unborn.

We realize that simply affirming our God-given responsibility to care for the earth will likely produce no tangible or effective results. Therefore, we pledge to find ways to curb ecological degradation through promoting biblical steward-ship habits and increasing awareness in our homes, businesses where we find influence, relationships with others and in our local churches. Many of our churches do not actively preach, promote or practice biblical creation care. We urge churches to begin doing so.

We realize that the primary impetus for prudent action must come from the will of the people, families and those in the private sector. Held to this standard of common good, action by government is often needed to assure the health and well-being of all people. We pledge, therefore, to give serious consideration to responsible policies that acceptably address the conditions set forth in this declaration.

CONCLUSION

We the undersigned, in accordance with our Christian moral convictions and Southern Baptist doctrines, pledge to act on the basis of the claims made in this document. We will not only teach the truths communicated here but also seek ways to implement the actions that follow from them. In the name of Jesus Christ our Lord, we urge all who read this declaration to join us in this effort. *Laus Deo!*

Climate Change

An Evangelical Call to Action

PREAMBLE

As American evangelical Christian leaders, we recognize both our opportunity and our responsibility to offer a biblically based moral witness that can help shape public policy in the most powerful nation on earth, and therefore contribute to the well-being of the entire world.[1] Whether we will enter the public square and offer our witness there is no longer an open question. We are in that square, and we will not withdraw.

We are proud of the evangelical community's long-standing commitment to the sanctity of human life. But we also offer moral witness in many venues and on many issues. Sometimes the issues that we have taken on, such as sex trafficking, genocide in the Sudan, and the AIDS epidemic in Africa, have surprised outside observers. While individuals and organizations can be called to concentrate on certain issues, we are not a single-issue movement. We seek to be true to our calling as Christian leaders, and above all faithful to Jesus Christ our Lord. Our attention, therefore, goes to whatever issues our faith requires us to address.

Over the last several years many of us have engaged in study, reflection, and prayer related to the issue of climate change (often called "global warming"). For most of us, until recently this has not been treated as a pressing issue or major priority. Indeed, many of us have required considerable convincing before becoming persuaded that climate change is a real problem and that it ought to matter to us as Christians. But now we have seen and heard enough to offer the following moral argument related to the matter of human-induced climate change. We commend the four simple but urgent claims offered in this document to all who will listen, beginning with our brothers and sisters in the Christian community, and urge all to take the appropriate actions that follow from them.

CLAIM I: HUMAN-INDUCED CLIMATE CHANGE IS REAL

Since 1995 there has been general agreement among those in the scientific community most seriously engaged with this issue that climate change is happening and is being caused mainly by human activities, especially the burning of fossil fuels. Evidence gathered since 1995 has only strengthened this conclusion.

Because all religious/moral claims about climate change are relevant only if climate change is real and is mainly human-induced, everything hinges on the scientific data. As evangelicals we have hesitated to speak on this issue until we could be more certain of the science of climate change, but the signatories now believe that the evidence demands action:

- The Intergovernmental Panel on Climate Change (IPCC), the world's most authoritative body of scientists and policy experts on the issue of global warming, has been studying this issue since the late 1980s. (From 1988–2002 the IPCC's assessment of the climate science was Chaired by Sir John Houghton, a devout evangelical Christian.) It has documented the steady rise in global temperatures over the last fifty years, projects that the average global temperature will continue to rise in the coming decades, and attributes "most of the warming" to human activities.
- The U.S. National Academy of Sciences, as well as all other G8 country scientific Academies (Great Britain, France, Germany, Japan, Canada, Italy, and Russia), has concurred with these judgments.
- In a 2004 report, and at the 2005 G8 summit, the Bush Administration has also acknowledged the reality of climate change and the likelihood that human activity is the cause of at least some of it.[2]

In the face of the breadth and depth of this scientific and governmental concern, only a small percentage of which is noted here, we are convinced that evangelicals must engage this issue without any further lingering over the basic reality of the problem or humanity's responsibility to address it.

CLAIM 2: THE CONSEQUENCES OF CLIMATE CHANGE WILL BE SIGNIFICANT, AND WILL HIT THE POOR THE HARDEST

The earth's natural systems are resilient but not infinitely so, and human civilizations are remarkably dependent on ecological stability and well-being. It is easy to forget this until that stability and well-being are threatened.

Even small rises in global temperatures will have such likely impacts as: sea level rise; more frequent heat waves, droughts, and extreme weather events such as torrential rains and floods; increased tropical diseases in now-temperate regions; and hurricanes that are more intense. It could lead to significant reduction in agricultural output, especially in poor countries. Low-lying regions, indeed entire islands, could find themselves under water. (This is not to mention the various negative impacts climate change could have on God's other creatures.)

Each of these impacts increases the likelihood of refugees from flooding or famine, violent conflicts, and international instability, which could lead to more security threats to our nation.

Poor nations and poor individuals have fewer resources available to cope with major challenges and threats. The consequences of global warming will therefore hit the poor the hardest, in part because those areas likely to be significantly affected first are in the poorest regions of the world. **Millions of people could die in this century because of climate change, most of them our poorest global neighbors.**

CLAIM 3: CHRISTIAN MORAL CONVICTIONS DEMAND OUR RESPONSE TO THE CLIMATE CHANGE PROBLEM

While we cannot here review the full range of relevant biblical convictions related to care of the creation, we emphasize the following points:

- Christians must care about climate change because we love God the Creator and Jesus our Lord, through whom and for whom the creation was made. This is God's world, and any damage that we do to God's world is an offense against God Himself (Gen. 1; Ps. 24; Col. 1:16).
- Christians must care about climate change because we are called to love our neighbors, to do unto others as we would have them do unto us, and to protect and care for the least of these as though each was Jesus Christ himself (Mt. 22:34–40; Mt. 7:12; Mt. 25:31–46).
- Christians, noting the fact that most of the climate change problem is human induced, are reminded that when God made humanity he commissioned us to exercise stewardship over the earth and its creatures. Climate change is the latest evidence of our failure to exercise proper stewardship, and constitutes a critical opportunity for us to do better (Gen. 1:26–28).

Love of God, love of neighbor, and the demands of stewardship are more than enough reason for evangelical Christians to respond to the climate change problem with moral passion and concrete action.

CLAIM 4: THE NEED TO ACT NOW IS URGENT. GOVERNMENTS, BUSINESSES, CHURCHES, AND INDIVIDUALS ALL HAVE A ROLE TO PLAY IN ADDRESSING CLIMATE CHANGE—STARTING NOW.

The basic task for all of the world's inhabitants is to find ways now to begin to reduce the carbon dioxide emissions from the burning of fossil fuels that are the primary cause of human-induced climate change.

There are several reasons for urgency. First, deadly impacts are being experienced now. Second, the oceans only warm slowly, creating a lag in experiencing the consequences. Much of the climate change to which we are already committed will not be realized for several decades. The consequences of the pollution we create now will be visited upon our children and grandchildren. Third, as individuals and as a society we are making long-term decisions today that will determine how much carbon dioxide we will emit in the future, such as whether to purchase energy efficient vehicles and appliances that will last for 10–20 years, or whether to build more coal-burning power plants that last for 50 years rather than investing more in energy efficiency and renewable energy.

In the United States, the most important immediate step that can be taken at the federal level is to pass and implement national legislation requiring sufficient economy-wide reductions in carbon dioxide emissions through cost-effective, market-based mechanisms such as a cap-and-trade program. On June 22, 2005 the Senate passed the Domenici-Bingaman resolution affirming this approach, and a number of major energy companies now acknowledge that this method is best both for the environment and for business.

We commend the Senators who have taken this stand and encourage them to fulfill their pledge. We also applaud the steps taken by such companies as BP, Shell, General Electric, Cinergy, Duke Energy, and DuPont, all of which have moved ahead of the pace of government action through innovative measures implemented within their companies in the U.S. and around the world. In so doing they have offered timely leadership.

Numerous positive actions to prevent and mitigate climate change are being implemented across our society by state and local governments, churches, smaller businesses, and individuals. These commendable efforts focus on such matters as energy efficiency, the use of renewable energy, low CO_2 emitting technologies, and the purchase of hybrid vehicles. These efforts can easily be shown to save money, save energy, reduce global warming pollution as well as air pollution that harm human health, and eventually pay for themselves. There is much more to be done, but these pioneers are already helping to show the way forward.

Finally, while we must reduce our global warming pollution to help mitigate the impacts of climate change, as a society and as individuals we must also help the poor adapt to the significant harm that global warming will cause.

CONCLUSION

We the undersigned pledge to act on the basis of the claims made in this document. We will not only teach the truths communicated here but also seek ways to implement the actions that follow from them. In the name of Jesus Christ our Lord, we urge all who read this declaration to join us in this effort.

NOTES

1. Cf. "For the Health of the Nation: An Evangelical Call to Civic Responsibility," approved by National Association of Evangelicals, October 8, 2004.

2. Intergovernmental Panel on Climate Change 2001, Summary for Policymakers; http://www.grida.no/climate/ipcc_tar/wg1/007.htm. (See also the main

IPCC website, www.ipcc.ch.) For the confirmation of the IPCC's findings from the U.S. National Academy of Sciences, see, *Climate Change Science: An Analysis of Some Key Questions* (2001); http://books.nap.edu/html/climatechange /summary.html. For the statement by the G8 Academies (plus those of Brazil, India, and China) see *Joint Science Academies Statement: Global Response to Climate Change,* (June 2005): http://nationalacademies.org/onpi/06072005 .pdf. Another major international report that confirms the IPCC's conclusions comes from the Arctic Climate Impact Assessment. See their *Impacts of a Warming Climate,* Cambridge University Press, November 2004, p.2; http:// amap.no/acia/. Another important statement is from the American Geophysical Union, "Human Impacts on Climate," December 2003, http://www.agu.org /sci_soc/policy/climate_change_position.html. For the Bush Administration's perspective, see *Our Changing Planet: The U.S. Climate Change Science Program for Fiscal Years 2004 and 2005,* p.47; http://www.usgcrp.gov/usgcrp /Library/ocp2004–5/default.htm. For the 2005 G8 statement, see http://www .number-10.gov.uk/output/Page7881.asp.

Notes

INTRODUCTION

1. VCY America Ministry, "About."

2. United States Congress Joint Committee on Printing, "2015–2016 Official Congressional Directory"; Terris, "Jim Inhofe"; Faith and Action in the Nation's Capital, "Senior Senator."

3. Inhofe, "Catastrophic Global Warming."

4. There has been considerable debate about what terminology to use when discussing those who doubt the scientific consensus on climate change. In this study, I am most interested in people who oppose action to mitigate climate change, so I have grouped them together under one heading: *climate skeptics*. In a departure from other studies, I therefore consider both people who dispute that the climate is changing at all and those who doubt whether such changes are caused primarily by human activities to be climate skeptics. I do so because the people I met during the course of my research who disagreed that anything needed to be done to stop or slow climate change held both positions. A second terminological concern comes from scientists, some of whom have opposed the use of the term *skeptic* to describe doubt or denial of anthropogenic climate change, on the grounds that it could be confused with the more rigorous practice of scientific skepticism (which includes debunking mysticism and pseudoscience, for example). Despite the validity of such concerns, I have still chosen to employ the term *skepticism* because I found it to be the most accurate description of my informants' tone and attitudes toward climate change, and because the laypeople and religious leaders I discuss are unlikely to be confused with practicing scientists.

5. This language comes from the New Living Translation, which the evangelical publisher Tyndale House first printed in 1996. In *The Greatest Hoax*, Inhofe adds that a daily devotional book published by the evangelical crusader

Bill Bright (founder of Campus Crusade for Christ) drew his attention to this line. He describes turning to the book, called *Promises*, "many times during my global warming fight" (70).

6. Tashman, "James Inhofe."

7. Funk and Alper, "Religion and Views on Climate."

8. Mills, Rabe, and Borick, "Acceptance of Global Warming."

9. Arbuckle and Konisky, "Role of Religion"; Kilburn, "Religion and Foundations."

10. McCright and Dunlap, "Politicization of Climate Change"; Egan and Mullin, "Climate Change"; Arbuckle and Konisky, "Role of Religion"; Kilburn, "Religion and Foundations"; Shao, "Weather, Climate, Politics, or God?"; Smith and Leiserowitz, "American Evangelicals"; Ecklund et al., "Examining Links"; Barker and Bearce, "End-Times Theology"; Arbuckle, "Interaction of Religion."

11. Ecklund et al., "Examining Links"; Arbuckle and Konisky, "Role of Religion."

12. Kilburn, "Religion and Foundations," 482.

13. According to the Pew Research Center, just 11 percent of white evangelicals fully accepted the theory of evolution as of 2014. See Pew Research Center, "U.S. Public Becoming Less Religious."

14. Evangelicalism is both a movement and a tradition. See Kellstedt and Green, "Knowing God's Many People," 57–58.

15. NAE, "What Is an Evangelical?" This definition derives from the work of the historian David Bebbington. See *Evangelicalism in Modern Britain*.

16. Williams, *God's Own Party*, 1.

17. African American evangelicals are often studied separately from white evangelicals for this reason. Memorably, one African American evangelical pastor I interviewed described himself as frequently being "the only chip in the cookie dough" at evangelical events.

18. Eskridge, "Defining the Term."

19. Pew Research Center, "Growing Numbers of Americans."

20. Pew Research Center, "Views about Abortion." On changing evangelical attitudes toward abortion, see Williams, *God's Own Party*, 111–120.

21. Pew Research Center, "Political Preference of U.S. Religious Groups."

22. Greeley and Hout, *Truth about Conservative Christians*, 1–2.

23. EEN, "Our Witness."

24. Wilkinson, *Between God and Green*, 25.

25. Harden, "Greening of Evangelicals."

26. Goodstein, "Evangelical Leaders."

27. Wilkinson, *Between God and Green*, xi.

28. Banerjee, "Southern Baptists."

29. Veldman et al., *How the World's Religions Are Responding*, 7. I owe this approach to Bron Taylor, who has argued convincingly of the need for an empirical turn in research on religion and the environment in order to disentangle "wishful thinking" about potential greening from on-the-ground reality. See Taylor, "Critical Perspectives" and "Toward a Robust Scientific Investigation."

30. The degree of empathy one should have toward one's informants is much debated. See McCutcheon, *Insider/Outsider Problem*.

31. Dunlap et al., "Measuring Endorsement."

32. This is according to the New International Version. When referencing the Bible for purposes of general discussion, I quote this version. I use the King James Version when discussing Bible quotations in the context of my informants' views because they typically preferred this version.

33. Wald, Owen, and Hill, "Churches as Political Communities."

34. Djupe and Hunt, "Beyond the Lynn White Thesis."

35. Moyers, "Welcome to Doomsday."

36. Pew Research Center, "Jesus Christ's Return to Earth."

37. Orr, "Armageddon versus Extinction," 291.

38. Boyd, "Christianity and the Environment"; Eckberg and Blocker, "Christianity, Environmentalism, and the Theoretical Problem of Fundamentalism"; Greeley, "Religion and Attitudes"; Guth et al., "Theological Perspectives" and "Faith and the Environment"; Hand and van Liere, "Religion, Mastery-over-Nature, and Environmental Concern"; Jones et al., *Believers, Sympathizers and Skeptics;* Sherkat and Ellison, "Structuring the Religion-Environment Connection"; Slimak and Dietz, "Personal Values, Beliefs, and Ecological Risk Perception."

39. Eschatology is an area of theological study that focuses on endings, both cosmic (the end of the world) and personal (death). See Hoekema, *Bible and the Future,* 1.

40. Marsden, *Fundamentalism and American Culture,* 6, 14.

41. LaHaye and Noebel, *Mind Siege,* 35.

42. Smith et al., *American Evangelicalism.*

43. Page, "Spirited and Christ-Like Climate Debate."

44. Ward, "Introduction," xviii.

45. Bean and Teles, "Spreading the Gospel," 2–3.

46. The scarcity of polls conducted prior to the ECI make it difficult to determine how much the Christian Right's campaign contributed to the phenomenon of evangelical climate skepticism, but it seems that skepticism was fairly well established by 2006. See chapter 7, note 9.

47. Dunlap and McCright, "Climate Change Denial."

48. Nagle, "Evangelical Debate"; Roberts, "Evangelicals and Climate Change"; Simmons, "Evangelical Environmentalism"; Wilkinson, "Climate's Salvation"; Wright, "Tearing Down the Green"; Hayhoe, "Climate, Politics and Religion"; Oreskes, "Religion and Climate Change Skepticism."

49. Kearns, "Cooking the Truth"; McCammack, "Hot Damned America"; Nagle, "Evangelical Debate"; Roberts, "Evangelicals and Climate Change"; Skeel, "Evangelicals, Climate Change, and Consumption"; Wilkinson, "Climate's Salvation" and *Between God and Green;* Wright, "Tearing Down the Green."

50. Curry, "Christians and Climate Change"; Haluza-DeLay, "Churches Engaging the Environment"; Kearns, "Cooking the Truth"; Nagle, "Evangelical Debate."

51. White, "Historical Roots," 1205.

52. Ibid.

53. See discussion in Guth et al., "Faith and the Environment," 367. For critical reviews of this literature, see Djupe and Hunt, "Beyond the Lynn White Thesis," and Taylor, Van Wieren, and Zaleha, "Lynn White, Jr."

54. Kearns, "Cooking the Truth"; McCammack, "Hot Damned America"; Simmons, "Evangelical Environmentalism"; Skeel, "Evangelicals, Climate Change, and Consumption."

55. Kearns, "Cooking the Truth"; Roberts, "Evangelicals and Climate Change"; Skeel, "Evangelicals, Climate Change, and Consumption"; DeWitt, "Creation's Environmental Challenge."

56. Curry, "Christians and Climate Change"; Fowler, *Greening of Protestant Thought;* Hulme, *Why We Disagree,* 155; Kearns, "Cooking the Truth"; Larsen, "God's Gardeners"; Maier, "Green Millennialism"; Nagle, "Evangelical Debate"; Roberts, "Evangelicals and Climate Change"; Simmons, "Evangelical Environmentalism"; Truesdale, "Last Things First"; Waskey, "Religion."

57. For more on the bias toward conceiving of religion as static in terms of its influence, as well as toward cognitive and doctrinal influences, see Djupe and Hunt, "Beyond the Lynn White Thesis."

58. Kahan et al., "Polarizing Impact of Science Literacy," 734.

59. Djupe and Hunt, "Beyond the Lynn White Thesis"; Jenkins, "After Lynn White"; Stoll, "Sinners in the Hands of an Ecologic Crisis."

60. Glaser and Strauss, *Discovery of Grounded Theory.*

61. Corbin and Strauss, "Grounded Theory Research," 11.

62. Krueger and Casey, *Focus Groups,* 8, 19.

63. Glacken, *Traces,* xiii.

64. Griffith, *God's Daughters.*

65. Ibid., 11, 12.

66. Ibid., 11.

67. Kellstedt and Smidt, "Measuring Fundamentalism"; Woodberry and Smith, "Fundamentalism et al."

68. The three main strategies for identifying white evangelicals are by denomination, beliefs, or self-identification. For more on these strategies, see Kellstedt and Green, "Knowing God's Many People"; Woodberry and Smith, "Fundamentalism et al."; and Smith et al., *American Evangelicalism,* appendix B.

69. See Woodberry and Smith, "Fundamentalism et al."

70. Woodberry and Smith's "Fundamentalism et al." is a good primer. For more detailed histories, see Marsden, *Fundamentalism and American Culture;* Sutton, *American Apocalypse;* and Carpenter, "Fundamentalist Institutions."

71. Marsden, *Understanding Fundamentalism and Evangelicalism,* 12.

72. As Sutton points out, the desire to expel liberals was driven not only by theological concerns but by financial ones. Sutton, *American Apocalypse,* 108.

73. Marsden, *Fundamentalism and American Culture,* 119.

74. Quoted in Sutton, *American Apocalypse,* 101.

75. Woodberry and Smith, "Fundamentalism et al.," 28.

76. Harding, "Representing Fundamentalism."

77. Woodberry and Smith, "Fundamentalism et al.," 26.

78. Speaking in tongues, also known as *glossolalia,* refers to speech or speech-like utterances produced during intense religious experiences. According to the New Testament, the phenomenon first occurred among Jesus's followers at Pentecost, which explains the Pentecostal movement's choice of name. See Wacker, *Heaven Below,* 5.

79. Woodberry and Smith, "Fundamentalism et al.," 30.

80. Green's other categories for evangelicals were "centrist" and "modernist." He constructed the scales as follows: "*Traditionalists* were those who claimed to be fundamentalist, evangelical, Pentecostal, or charismatic, or those without movement identification who agreed in preserving religious traditions. *Modernists* were those who claimed to be liberal or progressive, ecumenical or mainline and those without a movement identification who agreed in adopting modern religious beliefs and practices" (Green, "American Religious Landscape," 55–56; italics added).

81. Ibid., 4, 55–56.

82. In fact, many leaders in the Christian Right claim to speak for all evangelicals, rather than just those who support their agenda.

83. Green, "Fifth National Survey," 26.

84. Whitehead, "Christian Nationalism," 21.

85. See Connable, "'Christian Nation' Thesis"; see also chapter 4.

86. Harvey, "At Ease in Zion," 64–65.

87. Shibley, *Resurgent Evangelicalism.*

88. Woodberry and Smith, "Fundamentalism et al.," 32.

89. Shibley, *Resurgent Evangelicalism.*

CHAPTER ONE. THE END-TIME APATHY HYPOTHESIS

1. Pierson and Roe, "Mark Driscoll."

2. Ibid.; Merritt, "Is Mark Driscoll This Generation's Pat Robertson?"; see also Pyle, "Why Mark Driscoll's Theology of SUV's Matters."

3. PRRI and RNS, "Few Americans."

4. Cizik, "Natural Disasters."

5. Henneberger, "2000 Campaign"; Gore, *Earth in the Balance,* 263.

6. Wilson, *Creation,* 4, 6.

7. Nash, *Rights of Nature,* 92.

8. Kaufman and Franz, *Biosphere 2000,* 19.

9. Ibid., 19.

10. Numerous other sources could be named. See Hendricks, *Divine Destruction,* 20; Orr, "Armageddon versus Extinction," 290; Leduc, "Fueling America's Climatic Apocalypse," 276; Vanderheiden, *Atmospheric Justice,* 21–22; Hulme, *Why We Disagree,* 154–155; Waskey, "Religion"; Leggett, *Carbon War,* 173–175; Northcott, "Dominion Lie," 94; St. Clair, "Santorum"; Dolan, "Belief in Biblical End-Times"; Alleyne, "Maybe Religion Is the Answer"; Wood, "Senator Santorum's Planet."

11. Watt, *Caught in the Conflict,* 98.

12. Here is a sampling of the many texts that have cited Watt to illustrate the end-time apathy hypothesis: Davidson, "Dirty Secrets"; Fowler, *Greening of Protestant Thought,* 47; Guth et al., "Theological Perspectives," 377; Guth et al., "Faith and the Environment," 368; Haberman, *River of Love,* 231, n. 21; Harden, "Greening of Evangelicals," A01; Kearns, "Noah's Ark," 353; Kennedy, *Crimes against Nature,* 26; Rossing, *Rapture Exposed,* 7; Shabecoff, *Fierce Green Fire,* 203; Tucker, "Emerging Alliance," 4; Weber, *Apocalypses,* 202.

13. Leshy, "Unraveling the Sagebrush Rebellion," 319.

14. Hooper, "Reagan and the 'Sagebrush Rebellion,'" B2.

15. Schreibman, "Sierra Club Criticizes."

16. Arnold, *At the Eye of the Storm*, 22.

17. Ibid., 24–25.

18. Shabecoff, *Fierce Green Fire*, 208.

19. Kilpatrick, "Why Don't They Like James Watt?," A17; Pasztor, "James Watt Tackles."

20. Watt, *Caught in the Conflict*, 84.

21. Williams argues persuasively that evangelicals never opted out of politics. Nevertheless, they did not develop a strong partisan commitment until 1980 (Williams, *God's Own Party*, 2).

22. Although Reagan pioneered the strategy of seeking the evangelical vote, in the 1980 election evangelicals disproportionately voted for Jimmy Carter. See Woodberry and Smith, "Fundamentalism et al.," 44.

23. Watt, *Caught in the Conflict*, 97.

24. Dewar and Lyons, "All but One," A5.

25. Peterson, "Executive Notes." Leilani Watt notes that she and her husband "dismissed such stories for what they were—pure gossip with no basis in fact" (Watt, *Caught in the Conflict*, 97).

26. Omang, "Watt, a Man with a Mission," A1.

27. Ajemian, "Zealous Lord," 29.

28. Pasztor, "James Watt Tackles," 1.

29. Ibid.

30. Barkun, "Divided Apocalypse," 257–258.

31. Ibid., 261. The historian Paul Boyer offers essentially the same explanation for why end-time beliefs suddenly became a subject of scholarly interest (Boyer, *When Time Shall Be No More*, 16).

32. Martin, "Waiting for the End," 35.

33. Ibid., 36.

34. Schell, "Talk of the Town," 35. Schell was himself an apocalyptic, though of a different sort. Barkun argued that Schell's book *The Fate of the Earth* (1982) "turns the struggle to preserve life from extinction into a quasi-religious quest" (Barkun, "Divided Apocalypse," 264).

35. Wolf, "God, James Watt, and the Public's Land," 65.

36. Reed, "In the Matter of Mr. Watt," 6, 8.

37. Adams, "Stop Watt!," 14.

38. Pasztor, "James Watt Tackles"; Castelli, "What's Watt with God"; Castelli, "Environmental Gospel," 2; Kilian, "Hazy View," 8.

39. Arnold, *At the Eye of the Storm*, 76; Castelli, "Environmental Gospel," 2.

40. Wald and Calhoun-Brown, *Religion and Politics*, 224.

41. Ibid., 225.

42. Cited in Goldenberg, "Worst of Times."

43. Dunlap and McCright, "Organized Climate Change Denial," 152.

44. Goldenberg, "Worst of Times."

45. Scherer, "George Bush's War on Nature."

46. Scherer, "Why Ecocide Is 'Good News'" and "Religious Wrong." I originally located the latter article in E-*The Environmental Magazine*'s archives with a cover date of May/June 2003. It is now posted on the magazine's website as an editorial with the date July 20, 2004. Since the article opens by mentioning the opening of the 108th Congress, which occurred in 2003, I believe the actual publication date was 2003.

47. Scherer, "Christian-Right Views."

48. Boyer, *When Time Shall Be No More;* Wojcik, *End of the World;* Barkun, "Divided Apocalypse."

49. Email communication with Lisa Hymas, senior editor, Grist.org, April 14, 2015.

50. Scherer's and Phillips's arguments were not mutually exclusive. *American Theocracy* contains several passages suggesting that conservative Christian anti-environmentalism was rooted in end-time beliefs. Against scholars who dismissed such views as fringe, Phillips argued that "in the United States of the early 2000s . . . 'fringe' views can be held by 20 to 30 million people, which is more than a fringe—demographically at least" (Phillips, *American Theocracy*, 401, fn. 78).

51. Brinkley, "Clear and Present Dangers."

52. Moyers, "Welcome to Doomsday."

53. Strupp, "Bill Moyers Apologizes."

54. Moyers, "Welcome to Doomsday."

55. Moyers, *Welcome to Doomsday.*

56. *Miami Herald,* "Environment in Serious Trouble," 24A; Moyers, "With God on Their Side"; Moyers, "If We Don't Act Soon."

57. See, for example, Ayers, "Apocalypse Tomorrow."

58. Watt, *Caught in the Conflict,* 98.

59. Ibid.

60. A few scholars have cited the full quotation. See Curry-Roper, "Contemporary Christian Eschatologies," 162, 166; Nagle, "Evangelical Debate," 70.

61. Committee on Interior and Insular Affairs, *Reorganization of the Office of Surface Mining,* 23.

62. Watt, *Caught in the Conflict,* 100, 103.

63. Prochnau, "Watt Controversy," A1.

64. Larsen, "God's Gardeners," 170. The religious dimension of the tension between environmentalists and Watt is unmistakable. As a demonstrator at one of Watt's talks bluntly stated, "Down with God, Up with Trees" (Watt, *Caught in the Conflict,* 84). Leilani Watt recalled that "some of the signs so blasphemed God that I couldn't even repeat them" (ibid., 83). Recent work in religious studies has shown that Leilani was not wrong to characterize the disagreement as having religious dimensions. See Berry, *Devoted to Nature;* Taylor, *Dark Green Religion;* Stoll, *Inherit the Holy Mountain.*

65. One of the few that do not unravel first appeared in Scherer's *Grist* article, where he quoted the webmaster of RaptureReady.com, Todd Strandberg. In a response to Moyers (and, implicitly, to Scherer) posted on RaptureReady .com, Strandberg denied that he or any Christian group "actively call[ed] for

environmental destruction." Nevertheless, Strandberg concluded by asserting that "any preacher who decides to get involved in environmental issues is like a heart surgeon who suddenly leaves an operation to fix a clogged toilet" (Strandberg, "Bible Prophecy and Environmentalism"). This seems to validate Scherer's basic argument.

66. Worthen, "Who Would Jesus Smack Down?"; Driscoll, "Catalyst, Comedy and Critics."

67. Wilson was raised Southern Baptist (Wilson, *Consilience*, 6). Gore grew up alternating between attending Southern Baptist and Churches of Christ churches (Steinfels, "Beliefs"). Moyers was raised Southern Baptist (Allen, "How Bill Moyers Changed Baptist Journalism").

68. Other evangelicals who give credence to the end-time apathy hypothesis include Tony Campolo and Al Truesdale. See Campolo, *How to Rescue the Earth*, 92; Truesdale, "Last Things First," 116.

69. Watt, "Religious Left's Lies," A19.

70. Ibid. See also Beal, "Can a Premillennialist Consistently Entertain a Concern for the Environment?"

71. Ball, "Ungodly Distortions."

72. Guth et al., "Theological Perspectives."

73. Ibid., 380.

74. Guth et al., "Faith and the Environment," 364, 376.

75. Barker and Bearce, "End-Times Theology," 272.

76. Ibid.

77. Ibid., 275 (emphasis in original).

78. Smith, Hempel, and MacIlroy, "What's 'Evangelical' Got to Do with It?," 310–312. Both those affiliated with evangelical denominations and biblical literalists were less concerned about future environmental problems than their mainline counterparts. The future environmental problems tested were whether climate change will have disastrous effects, whether humans will exhaust the earth's natural resources, and whether humans will destroy most of the plant and animal life on earth (neither evangelical affiliation nor biblical literalism had a statistically significant relationship with the last).

79. Weber, "Millennialism," 367–369.

80. Ibid., 368.

81. Ibid., 367.

82. Clouse, Hosack, and Pierard, *New Millennium Manual*, 55.

83. Theologically liberal postmillennialists do not believe that Jesus will literally return to earth; they define the kingdom of God in terms of ethics. Weber, "Millennialism," 378.

84. Sutton, *American Apocalypse*, 27.

85. Boyer, *When Time Shall Be No More*, 117, 162–163, 187–189, 333; Barkun, "Divided Apocalypse."

86. Barkun, "Divided Apocalypse," 260.

87. Sutton, *American Apocalypse*, 346.

88. Pew Research Center, "Jesus Christ's Return to Earth."

89. I did not include postmillennialists because, for historical reasons, evangelical postmillennialists are few in number. Christian Reconstructionists are an

important exception, but lay followers are difficult to identify. The movement has, however, been influential in encouraging Christians (and leaders in the Christian Right) to set aside the premillennialist tendency to withdraw from politics in favor of more active involvement in public life (Clarkson, "Christian Reconstructionists," 86). Hence, postmillennialism may have shaped my informants' outlook indirectly.

90. Cizik, "Natural Disasters."

CHAPTER TWO. PRACTICAL ENVIRONMENTALISM

1. McKibben, "Link."
2. Severson and Johnson, "Drought Spreads Pain."
3. Botelho, "Weeks-Long Wildfire."
4. Janisse Ray has beautifully illustrated this point in her autobiographical books *Ecology of a Cracker Childhood* (1999) and *Wild Card Quilt: The Ecology of Home* (2003).
5. Environmentalism has tended to address the concerns of urbanites and, while evidence is spotty, seems to appeal particularly to those who have moved away from traditional forms of Christianity. See Fox, *John Muir and His Legacy,* ch. 11; Guth et al., "Faith and the Environment," 377; Shaiko, "Religion, Politics, and Environmental Concern"; Shibley, "Sacred Nature."
6. Leopold, *Sand County Almanac,* 197.
7. Orr, *Ecological Literacy,* 86.
8. Thomashow, *Bringing the Biosphere Home,* 5.
9. Ibid., 5.
10. Norgaard, *Living in Denial.*
11. Research & Polling Inc., "Focus Group Report." I thank Gregory Green for sharing this report with me.
12. Jones, "In U.S., Concern about Environmental Threats Eases"; Pew Research Center, "Protecting the Environment."
13. Greeley and Hout, *Truth about Conservative Christians,* 14.
14. The term *cultural tool kit* comes from Swidler, "Culture in Action."
15. Stark and Glock, "Prejudice and the Churches," 80–81; cited in Emerson and Smith, *Divided by Faith,* 77.
16. Emerson and Smith, *Divided by Faith,* 78.
17. Ibid., 79.
18. The emphasis on accountable free-will individualism is rooted in a subcultural discourse specific to American evangelicals. See Bean, *Politics of Evangelical Identity,* ch. 5.
19. Dryzek, *Politics of the Earth,* 21.
20. Dunlap et al., "Measuring Endorsement," 426.
21. See ibid., 433.
22. Wilkinson, *Between God and Green,* 105–108.
23. Palmer, "Overview of Environmental Ethics," 8.
24. White, "Historical Roots," 1205.
25. Religious environmental organizations seem to have accommodated this apparently widespread anthropocentric bent. In his interviews with a cross

section of sixty such organizations, many of them Christian, the sociologist Stephen Ellingson observed that "they tend to work on ecological issues that have a clear impact on human beings" (Ellingson, *To Care for Creation*, 3).

26. Jones and Dunlap, "Social Bases of Environmental Concern."

27. One of Emerson and Smith's African American informants used almost identical language to dismiss the problem of race, saying, "We don't have a race problem, we have a sin problem" (Emerson and Smith, *Divided by Faith*, 78). As they point out, for evangelicals sin is fundamentally an individual concept, and as such it tends to draw attention away from (or to obscure) problems that are structural in nature.

28. Dryzek, *Politics of the Earth*, 12–13.

CHAPTER THREE. END-TIME BELIEFS AND CLIMATE CHANGE

1. I adapt the terms *hot* and *cool* millennialism from the religious studies scholar Scott Lowe. In his work on Chinese millennialist movements, he has distinguished between "cool millennialists" who believed that the millennium was far off and "hot millennialists" who hoped to speed its arrival (Lowe, "Chinese Millennial Movements," 318).

2. The other two focus groups that had hot millennialist participants were those held at the Church of Christ and at the Missionary Baptist church. Although it would be unwise to generalize from such a small sample, by this count hot millennialists did seem to be concentrated in premillennialist denominations.

3. Poloma and Green, *Assemblies of God*, 5.

4. Assemblies of God, "End-Time Events."

5. Assemblies of God, "Our Core Doctrines."

6. Lawson, "Persistence of Apocalypticism," 215.

7. Ibid., 216.

8. 2 Peter 3:12.

9. Boyer, *When Time Shall Be No More*, 117.

10. For example, in 2008 the Jehovah's Witnesses posted an article about global warming titled "Earth Has a 'Fever.' Is There a Cure?" on their website. The article cited the finding of the IPCC's Fourth Assessment Report (released in 2007) that "the warming of the climate system is unequivocal" and argued that this "global problem" is in need of a "global solution"—the restoration of God's kingdom. Continuing his evident power over nature, Jesus will "recreate or re-new conditions on earth" when he returns. A 2014 article argued similarly that "the Bible foretold a time when man would 'ruin *the earth*' (Revelation 11:18)" but reassured readers that "God will rebalance the ecological budget" (Jehovah's Witnesses, "Will Man Ruin the Earth").

11. Interjections from other participants have been deleted for clarity.

12. Feldman et al., "Climate on Cable."

13. Maibach et al., "Global Warming's Six Americas," 4.

14. Ibid., 4.

15. The devastating Tohoku earthquake and tsunami that hit Japan in March 2011 were probably fresh in Roger's mind.

16. An important study that argues along these lines is Michael Barkun's *Disaster and the Millennium* (1974).

17. Land, "Richard Land Extended Interview."

18. Thompson, *Waiting for Antichrist,* 9–10.

CHAPTER FOUR. THE EMBATTLED MENTALITY AND CLIMATE SKEPTICISM

1. Mirroring Joshua's description, on the denomination's website it describes itself as "undenominational" (Churches of Christ, "Internet Ministries").

2. Miller, *Silencing of God.*

3. Because I relied on group discussions as my primary means of data collection, I do not have precise numbers of climate change accepters versus skeptics. Still, it was clear that skepticism predominated. Out of fifty-three participants, only three individuals affirmed during the course of the discussion that they believed climate change was occurring, that it was mostly human caused, and that this was a problem (three others expressed vague or ambiguous support for this position). The consensus was similar among pastors: only one of fifteen accepted that the climate was changing as a result of human activities.

4. The premillennialism of Seventh-day Adventists differs from that typically taught in evangelical circles. However, given that both believe in a period of worsening conditions prior to Jesus's return (conditions that could include environmental decline), the end-time apathy hypothesis should still, in theory, explain their environmental attitudes.

5. Smith et al., *American Evangelicalism,* 89.

6. PRRI and Brookings Institute, "How Immigration and Concerns," 17.

7. Hawkins, "7 Examples"; Unruh, "Attacks."

8. Regarding whether this questioning route would have primed my informants to respond to my later questions about the environment through the lens of embattlement, thereby inadvertently biasing the results, the congruence between my findings and those of other researchers who used different techniques (discussed in chapter 6) suggests not.

9. Smith et al., *American Evangelicalism,* 126.

10. Barack Obama's 2008 presidential campaign featured "Hope and Change" as a slogan.

11. Ertelt, "Super Bowl Ad Gets Focus."

12. See Smith et al., *American Evangelicalism,* ch. 5.

13. Ibid., 144–145.

14. Quoted in Roberts, *Darwinism and the Divine,* 64.

15. Sutton, *American Apocalypse,* 141.

16. Ibid., 150.

17. Marsden, *Understanding Fundamentalism and Evangelicalism,* 14.

18. Sutton, *American Apocalypse,* 150.

19. Ibid., 157.

20. Quoted in ibid., 167 (emphasis in original).

21. Kruse, *One Nation under God,* 18.

22. Quoted in Sutton, *American Apocalypse,* 258.

23. Ibid., 287–288.
24. Quoted in ibid., 287.
25. Delfattore, *Fourth R,* 61.
26. Kruse, *One Nation under God,* 200.
27. Cited in ibid., 200.
28. Williams, *God's Own Party,* 67.
29. Ibid., 134.
30. Ibid., 136.
31. Ibid., 137–143.
32. Schaeffer, *Pollution and the Death of Man,* 77.
33. Ibid., 81.
34. Jackson and Perkins, *Personal Faith, Public Policy,* 17.
35. Balmer, *Evangelicalism in America,* 111–131.
36. Castelli, "Persecution Complexes," 160.
37. Williams, "Politicized Evangelicalism," 159–160.
38. LaHaye and Noebel, *Mind Siege,* 35. Williams credits LaHaye with popularizing the term *secular humanism* among conservative Protestants (Williams, "Politicized Evangelicalism," 160).
39. Williams, "Politicized Evangelicalism," 159. Lienesch provides a useful summary of how these views are tied to a specific reading of history (Lienesch, *Redeeming America,* 139–194).
40. Connable, "'Christian Nation' Thesis," 185.
41. Williams, "Politicized Evangelicalism," 151.
42. Bean, *Politics of Evangelical Identity,* 79.
43. Williams, "Politicized Evangelicalism," 158–159.
44. Ibid., 152.
45. Pastor Richard was not wrong to associate environmentalism with Hollywood, as a number of celebrities support environmental causes (Leonardo DiCaprio is perhaps the best known). But environmentalism did not originate in Hollywood, as he seemed to suggest.
46. They were not necessarily wrong to view environmentalists as cultural opponents, but few environmentalists would agree that their primary goal was to secularize the nation, so my informants' interpretations of the situation differed.
47. Matthew 6:34.
48. Glaser and Strauss, *Discovery of Grounded Theory,* 5.
49. Ibid., 5.
50. Egan and Mullin, "Climate Change: US Public Opinion," 216.
51. Dunlap and McCright, "Widening Gap," 28–29.
52. Begley et al., "Truth about Denial."
53. Dunlap and McCright, "Organized Climate Change Denial," 144.
54. Oreskes and Conway, *Merchants of Doubt.*
55. Dunlap and McCright, "Organized Climate Change Denial," 149.
56. Ibid.
57. Ibid., 150.
58. Cohen and Bell, "Congressional Insiders Poll," 6.
59. McDonald, "Changing Climate, Changing Minds," 46–48.

60. Robert Brulle and his colleagues' analysis of climate change attitudes from 2002 to 2010 showed that elite cues were a major factor driving shifts in opinion (Brulle, Carmichael, and Jenkins, "Shifting Public Opinion," 182). As for why Republican political elites are more likely to be skeptical in the first place, the Harvard sociologist Theda Skocpol suggests that while Republican politicians "no doubt had longstanding, practical reasons to listen to friendly business interests and oppose liberal environmentalists," the climate denial machine's disinformation campaign "gave them additional rationales for foot-dragging, as long as the science could be called 'unsettled'" (Skocpol, "Naming the Problem," 68–69; this report is no longer available online).

61. Brulle, Carmichael, and Jenkins, "Shifting Public Opinion," 182.

62. McCright, "Political Orientation," 246.

63. Feldman et al., "Climate on Cable," 3.

64. Cook et al., "Consensus on Consensus."

65. Elsasser and Dunlap, "Leading Voices," 763.

66. Knight and Greenberg, "Talk of the Enemy," 324.

67. Fine, "Notorious Support," 406.

68. Lahsen, "Climategate," 547–548.

69. Dunlap and McCright, "Climate Change Denial," 249. See also Knight and Greenberg, "Talk of the Enemy."

70. It is true that Gore has invested in the alternative energy sector. His explanation was that "I absolutely believe in investing in accordance with my beliefs and values" (Allen, "Al Gore 'Profiting'").

71. Knight and Greenberg, "Talk of the Enemy," 332.

72. It is notable that Kelly referenced control over education, since removing the Bible from public school classrooms was one of the key ways in which evangelicals felt that the federal government had unnecessarily and harmfully intervened.

73. Crouch, "Environmental Wager"; Ecklund et al., "Examining Links"; Kearns, "Cooking the Truth"; McCammack, "Hot Damned America"; Roberts, "Evangelicals and Climate Change"; Wilkinson, Between God and Green.

74. Wilkinson, Between God and Green, 95; Gauchat, "Politicization of Science."

75. In light of these parallels, it is interesting to note that some creationist organizations and global warming denialists have begun working together. In addition to chapter 7, see Roberts, "Evangelicals and Climate Change," and Stewart, "New Anti-science Assault."

76. I would argue that accepting "microevolution" does not amount to partially accepting Darwin's theory of evolution, since it was known that species changed between generations before Darwin's time. It is instructive, however, that my informants made this distinction.

77. Miller, Silencing of God.

78. Despite Miller's reference to Jesus's return, as with my informants, the embattled mentality seems to have had the stronger influence on his views about climate change. The website for Apologetics Press, where Miller is executive editor, contains an article about global warming in its section on end times. Rather than seeing global warming as a sign of the nearing end, it repeats

standard denialist talking points when it comes to global warming (Lyons, "Global Warming"). In another indication that Miller does not fit the end-time apathy hypothesis, he rejects premillennialism (Miller, "Is the Kingdom Yet to Be Established?").

CHAPTER FIVE. HOW EVANGELICAL SUBCULTURAL IDENTITY SUSTAINS CLIMATE SKEPTICISM

1. Despite its strong legacy in the social scientific literature, Matthew Riley argues persuasively that in White's view it was not simply anthropocentrism but democracy that was responsible for the ecological crisis. See Riley, "Democratic Roots," 942–945.

2. Zerubavel, *Social Mindscapes*, 9, 15.

3. Ibid., 39.

4. Grudem, *Systematic Theology*, 492.

5. It is not necessarily the case that scripture is silent about the climate or climate change. My informants could have cited Genesis 1:22 (as does Senator Inhofe, as described in the opening pages of this book). What seemed to be missing was the *tradition* of turning to this particular passage for guidance.

6. Williams, "Politicized Evangelicalism," 146–147.

7. Lichterman, "Religion and the Construction of Civic Identity."

8. Ibid., 86.

9. Ethan's remark calls to mind—and is likely indirectly linked to—the long-standing debate in natural resource management between "conservation" and "preservation" (see Callicott, "Brief History").

10. Horton, "Green Distinctions."

11. Smith et al., *American Evangelicalism*, 126.

12. Ibid., 126.

13. This finding supports Dan Kahan's "cultural cognition thesis," which states that "individuals, as a result of a complex of psychological mechanisms, tend to form perceptions of societal risks that cohere with values characteristic of groups with which they identify" (Kahan et al., "Polarizing Impact," 732). As he and his coauthors point out, climate change attitudes are strongly affected by such factors.

14. Steed, "Joke-Making Jews," 9.

15. Underlining this sense of social risk, Jill, my key informant who had lived in the area for decades, told me that when she brought up the idea of human-caused climate change with people who lived in the area, her comments were met with silence—a powerful means of communicating disapproval.

16. See, for example, Kellstedt and Green, "Knowing God's Many People."

17. Seventh-day Adventist Church, "28 Fundamental Beliefs." The page I originally accessed is no longer available, but the same quotation can be found in the 2015 document "28 Fundamental Beliefs," available at www.adventist .org/fileadmin/adventist.org/files/articles/official-statements/28Beliefs-Web.pdf.

18. Buettner, *Blue Zones*.

19. Stark and Bainbridge, *Future of Religion*, 49–51.

20. Ibid., ch. 7. There is an apparent conflict between church-sect theory, which predicts that when sects are successful they encounter pressure to reduce tension with society, and Smith and his colleagues' subcultural identity theory (presented in *American Evangelicalism: Embattled and Thriving*), which posits evangelicalism's state of high tension as the primary reason it is successful. The solution may be related to evangelicalism's size and historical legacy, which together have enabled it to heighten tension with less risk.

21. Lawson, "Persistence of Apocalypticism."

22. Ibid., 214.

23. Seventh-day Adventist Church, "Stewardship of the Environment."

24. Ibid.

25. Seventh-day Adventist Church, "Dangers of Climate Change."

26. Veldman, Szasz, and Haluza-DeLay, "Introduction: Climate Change and Religion," 261.

27. Seventh-day Adventist Church, "Glimpses of Our God," 64 (the pamphlet can be accessed at http://ssnet.org/lessons/12a/less08.html).

28. Ibid., 65.

29. Seventh-day Adventist Church, "Environment"; "Glimpses of Our God," 66.

30. Matthew 22:39. The guide references the New King James Version of the Bible, and this section follows suit.

31. Seventh-day Adventist Church, "Glimpses of Our God," 67–70.

32. Ibid., 70.

33. NAE, "Ecology." See also NAE, "Environment and Ecology" and "Stewardship."

34. NAE, "Caring for God's Creation."

35. Despite this reputation for conservatism, as Ammerman explored in great detail, there have been long-standing battles between "progressives" (or "moderates") and "fundamentalists" within the SBC (Ammerman, *Baptist Battles*).

CHAPTER SIX. SALT AND LIGHT

Adapted from Veldman, "What Is the Meaning of Greening?"

1. SBC, "On Global Warming."

2. Ibid.

3. SBECI, "Southern Baptist Declaration."

4. Banerjee, "Southern Baptists Back a Shift."

5. See, for example, Harden, "Greening of Evangelicals," A01.

6. As Mikko Lehtonen has written, "Meaning is never written as if readymade inside the text, but is formed within the reading of the text, which is affected by the position of the reader in contexts and cultural practices in addition to the text itself" (Lehtonen, *Cultural Analysis of Texts*, 73).

7. Jonathan Merritt, interview with Andrew Szasz, February 4, 2011. This quotation comes courtesy of an interview that Szasz graciously shared with me. All other quotations are from my interview with Jonathan Merritt, conducted on August 10, 2012, unless otherwise noted.

8. Jonathan Merritt, author interview, August 10, 2012. Jonathan did not tell me their official positions, but they are listed on the masthead of the fall 2007 (no. 34) edition of *Creation Care: A Christian Environmental Quarterly,* which the EEN used to publish. An archived version of the page (http://creationcare.org/media.php?what=21&c_id=5) can be found on the Internet Archive Wayback Machine.

9. ECI, "Climate Change: An Evangelical Call to Action."

10. Jonathan Merritt, author interview, August 10, 2012.

11. According to Jonathan Merritt, the co-drafters were James Merritt; Timothy George, who had signed the Evangelical Climate Initiative; Danny Akin, the president of the seminary Jonathan was then attending, Southeastern Baptist Theological Seminary; David Clark, another ECI signatory and president of Palm Beach Atlantic University; and David Dockery, president of Union University (author interview, August 10, 2012).

12. Of those listed on the website associated with the project (just 427), one hundred were pastors, twenty-seven worked in a leadership capacity in the Christian community (serving, for example, as presidents of Christian universities or as leaders of parachurch organizations), and twelve were professors. The majority of the remaining signatories were lay supporters. While the inclusion of high-profile leaders was noteworthy, the SBC claims to represent a network of fifty thousand churches, which makes 750 signatories a small proportion. The SBECI's website was www.baptistcreationcare.org/node/1, which is no longer available. According to the Internet Archive Wayback Machine, the site became inactive at the end of 2016.

13. Jonathan Merritt, author interview, August 10, 2012.

14. Banerjee, "Southern Baptists Back a Shift."

15. Hagerty, "Climate Change Prompts Debate"; Van Biema, "Greening of the Baptists"; CNN, "Southern Baptist Leaders Shift Position"; Lampman, "Southern Baptist Leaders Urge Climate Change Action."

16. Zoll, "Southern Baptist Leaders Take Unusual Step." The number of local news stories comes from an analysis of stories indexed in the database Access World News, which includes 5,419 news sources in the United States.

17. See Bergmann, "Climate Change Changes Religions"; Harrington, "Evangelicalism, Environmental Activism, and Climate Change"; Kearns, "Religious Climate Activism"; Wilkinson, *Between God and Green.*

18. When I asked Jonathan if he had ever considered leaving climate change out of the Declaration, he replied, "No I didn't, because I believe the problems that we are seeing with the climate are serious." Elsewhere he has described his stance as "middle-of-the-road," likely because he eschews apocalyptic predictions, while still accepting that humans have contributed something to the problem (Merritt, *Green Like God,* 162).

19. Radway, *Reading the Romance,* 11.

20. Lehtonen, *Cultural Analysis of Texts,* 144.

21. ASARB, "Southern Baptist Convention States."

22. Banerjee, "Southern Baptists Back a Shift"; Zoll, "Southern Baptist Leaders Take Unusual Step."

23. Van Biema, "Greening of the Baptists."

24. CNN, "Southern Baptist Leaders Shift Position."

25. Hagerty, "Climate Change Prompts Debate"; Van Biema, "Greening of the Baptists"; Lampman, "Southern Baptist Leaders Urge Climate Change Action."

26. The *Christian Science Monitor* story was the exception. See Lampman, "Southern Baptist Leaders Urge Climate Change Action."

27. Banerjee, "Southern Baptists Back a Shift."

28. Van Biema, "Greening of the Baptists."

29. Lampman, "Southern Baptist Leaders Urge Climate Change Action."

30. Lehtonen, *Cultural Analysis of Texts,* 126.

31. See, for example, Wardekker et al., "Ethics and Public Perception"; McKibben, "Gospel of Green"; and Brinton, "Green, Meet God," 12A.

32. Oreskes, "Scientific Consensus on Climate Change."

33. Harris, "Call to Good Stewardship."

34. Westbury, "Seminary President"; Page, "Frank Page Statement."

35. Westbury, "Younger Conservative Leaders."

36. Foust, "SBC Leaders."

37. Westbury, "Jonathan Merritt" (emphasis mine).

38. Page, "Spirited and Christ-Like Climate Debate."

39. Westbury, "Younger Conservative Leaders."

40. Westbury, "Jonathan Merritt."

41. Westbury, "Seminary President."

42. Harris, "Call to Good Stewardship."

43. ERLC Staff, "Calling Southern Baptists."

44. Westbury, "Seminary President" (parentheses in original).

45. Merritt, *Green Like God,* 85.

46. Westbury, "Younger Conservative Leaders."

47. Emphasis mine.

48. Emphasis mine.

49. SBC, "Resolution on Environmental Stewardship."

50. SBC, "On Environmentalism and Evangelicals." This resolution was presumably a response to the ECI, which had been announced five months earlier.

51. Smith et al., *American Evangelicalism,* 113.

52. See also Williams, "Politicized Evangelicalism"; Connable, "'Christian Nation' Thesis."

53. Wilkinson, *Between God and Green,* 100.

54. Carr, "Faithful Skeptics: Conservative," 140.

55. Carr et al., "Faithful Skeptics: Evangelical," 293.

56. Peifer et al., "How Evangelicals from Two Churches," 385.

57. Ibid.

58. Zaleha, "Tale of Two Christianities," 120.

59. Ibid., 96, 129, 127.

60. Roser-Renouf et al., *Global Warming, God, and the "End Times."* This survey also found that 24 percent of evangelical or born-again Christians thought global warming was a sign of the end times, and that 26 percent thought the end times were coming—and therefore there was no need to worry about global warming. Both statements could reflect the embattled mentality.

Respondents could have interpreted the first statement to mean that "the pro-motion of global warming by liberal/secular activists" is a sign of the end times. This interpretation has in fact been promoted by Tim LaHaye (see Ronan, "American Evangelicalism"). The second statement could be interpreted to mean "we do not need to worry about climate change because God is in control of the end times," which was a common sentiment among my informants. It is also important to note that the percentages given reflect all evangelicals, regard-less of race, perhaps understating the prevalence of such beliefs among white evangelicals.

61. Pew Research Center, "Tea Party and Religion."

62. Rabe and Borick, *National Surveys*. I thank Christopher Borick and Sarah Banas Mills for sharing these numbers with me. The figures given reflect percentages of the entire evangelical population sampled, regardless of race.

63. Non-evangelical Protestants and Catholics who cited religious reasons for climate skepticism, while doing so in smaller numbers, offered similar expla-nations for the years I examined, between 2010 and 2015. This finding merits further research.

64. Williams, *God's Own Party*, 111–120.

CHAPTER SEVEN. PREACHING THE GOSPEL
OF CLIMATE SKEPTICISM

1. Davenport, "As Climate Risks Rise," A1.

2. Stonestreet, "Paris Accord."

3. Ward, "Appendix B," 267.

4. Heartland Institute, "Who We Are."

5. Hurd, "World Leaders."

6. Ward, "Introduction," xviii.

7. Dunn and Tyler's study of four program genres in the evangelical mass media (advocacy talk shows, lifestyle talk shows, apologetics programs, and teaching programs) concluded that "most speakers and hosts avoided making politics the dominant theme." When such programs did address politics, how-ever, they generally "dismissed the idea that Christians could legitimately hold political views that differed from conservative orthodoxy." See Dunn and Tyler, "Resisting Pluralism," 146–147.

8. Ibid., 139.

9. A poll conducted in 2004, for example, found that only 40 percent of white evangelicals said that it was extremely important or very important for American foreign policy to combat global warming, compared to 53 percent of the general population (Religion and Ethics Newsweekly and U.S. News and World Report, "America's Evangelicals"). Similarly, an October 2005 poll found that only 33 percent of born-again or evangelical Christians designated dealing with global climate change as a top foreign policy goal for the United States, compared to 58 percent of non-born-again Christians (Pew Research Center, "America's Place in the World IV"). Another poll, from June 2005, found that 33 percent of evangelicals thought nothing should be done "until we are sure that global warming really is a problem," compared to just 17 percent

of non-evangelicals (PIPA, "Americans on Addressing World Poverty"; about 30 percent of both evangelicals and non-evangelicals in this survey thought global warming was a serious and pressing problem, which suggests that even prior to the ECI, evangelicals differed from the general public mainly in that they had a more active skeptical wing). A 2006 survey, conducted a month after the ECI was announced, similarly found "little resonance for that statement [the ECI] among white evangelical Protestants. They're less likely than others to see global warming as a threat to the global environment or to say the government should address it" (ABC News, Time Magazine, and Stanford University, "Intensity Spikes"). The only survey I have seen that presents a different picture is one funded by the EEN. Based on a nationally representative poll of one thousand born-again or evangelical Protestants conducted in September 2005, this survey found that 63 percent of born-again or evangelical Protestants agreed that global warming was an issue that needed to be addressed immediately (Ellison Research, "Nationwide Study"). It is unclear why this survey attained such different results from those listed above.

10. Ward, "Give the Winds," 129.

11. For a description of LaHaye's efforts, see Ronan, "American Evangelicalism."

12. Heltzel, *Jesus and Justice*, 146. In fact, according to Haggard, their prominence is the reason that evangelicals are so often conflated with the Religious Right.

13. Schlafly, who died in 2016, dedicated a number of columns and radio programs to debunking climate change (see Eagle Forum, "Global Warming"). As chapter 8 mentions, Weyrich signed some of the Cornwall Alliance's documents and penned at least one opinion piece advocating skepticism about anthropogenic climate change (Weyrich, "Falst [sic] Frenzy").

14. Colson discussed climate change just twice in online editorials penned in 2005. He posted only two more stories in 2006 and 2007 (if archives were complete) but wrote eight in 2008. *Answers with Ken Ham* criticized environmentalism without discussing climate change in 2005 and 2006, in a total of four episodes; the feature criticized climate science just once in 2007. In 2008, however, ten episodes promoted climate skepticism. CBN's older stories tended to be undated, making it difficult to assess its coverage of climate change, but I could date all the undated stories except two to 2006 or later on the basis of their content. This suggests that, like the others I reviewed, it covered the issue infrequently prior to the ECI. Swanson's *Generations Radio* had its first global warming story in March 2006, whereas Movieguide.org averaged one story per year between 2002 and 2008 but posted eleven in 2009. Other programs I reviewed did not have archives or had incomplete archives for the period before the ECI.

15. Beverly LaHaye's Concerned Women for America and the American Family Association are two organizations that were beyond the scope of this project but deserve further investigation.

16. The *Christian Post* is an important exception. My analysis of its news coverage in 2009 found that it posted slightly more accepting stories than skeptical ones. Opinion pieces were uniformly skeptical.

17. Conservative pundits have also played a role in advancing religious objections to climate change. For example, in 2013, Rush Limbaugh stated on his radio program *The Rush Limbaugh Show* that "if you believe in God, then intellectually you cannot believe in manmade global warming. You must be either agnostic or atheistic to believe that man controls something that he can't create" (Edwards, "Limbaugh"). E. Calvin Beisner, the founder and national spokesman of the Cornwall Alliance, also appeared on Glenn Beck's Fox News program in 2010, exemplifying the substantial crossover between secular and religious conservative circles. See also Coulter, *Godless*, 18–19.

18. Williams, *God's Own Party*, 271; Ward, "Introduction," xvii.

19. Kennedy, *Crimes against Nature*, 28–29.

20. CBN, "Al Gore."

21. Wilson, "Pat Robertson's Sweaty Global Warming Epiphany."

22. Legum, "Pat Robertson"; Banks, "Some Evangelicals Go Green."

23. Loller, "Gore Announces." The advertisement with Robertson can be viewed at www.youtube.com/watch?v=NhmpsUMdTH8 (April 17, 2008).

24. *CBN News*, "Team Effort."

25. Transcripts courtesy of Chris Roslan, Roslan and Campion Public Relations. Notably, this statement also contradicts the end-time apathy hypothesis.

26. Email communication, June 2, 2016.

27. The report can be viewed at www.youtube.com/watch?v=ZE0Tlg6BQsw (accessed May 30, 2016).

28. CBN, "CBN NewsWatch."

29. Ward, "Appendix B," 267.

30. Estimated via a SEMRush analysis conducted July 28, 2017. As of 2017, CBN self-reported 3.4 million unique visits during an average month (CBN, "Advertise with CBN").

31. Rank is according to Alexa.com's "Top 500 Sites on the Web" rankings (subcategory "Society," subcategory "Religion and Spirituality," subcategory "Christianity"). Accessed May 31, 2016.

32. BreakPoint, "About."

33. Lindsay, *Faith in the Halls*, 59.

34. BreakPoint, "About."

35. Colson, "BreakPoint: 'Not Evil, Just Wrong.'" While Colson described the film as being released by the Cornwall Alliance, I have found no evidence that the Cornwall Alliance was responsible for it. The Internet Archive Wayback Machine's October 18, 2009, web-crawl of the Cornwall Alliance's website shows that it did endorse the film, however.

36. Colson, "3 Reasons." Implying that environmentalists are latter-day communists has been a common tactic employed by leaders in the Christian Right.

37. Colson, "BreakPoint: Global Warming as Religion."

38. Search conducted May 24, 2016, using the archive at www.breakpoint.org/bpcommentaries/breakpoint-commentaries-archive (page no longer available).

39. This analysis was based on a search of the *BreakPoint* archives conducted in August 2016.

40. Colson, "BreakPoint: Global Warming and Ideology."

41. Estimated via a SEMRush analysis conducted July 28, 2017. The "*Christian Post* Media Kit" reports that the website has ten million unique monthly visitors.

42. Alexa, "Top 500 Sites on the Web" (category "Society," subcategory "Religion and Spirituality").

43. Salem Web Network, "Advertising."

44. Colson, "Global Warming as Religion" and "Engineering the Earth."

45. Banerjee, "Rev. D. James Kennedy."

46. Religion News Service, "Posthumous Preaching."

47. Kennedy and Beisner, *Overheated*, 28.

48. TruthsThatTransform.org (home page).

49. Kennedy and Newcombe, *How Would Jesus Vote?*, 133.

50. Ibid., 134.

51. Coral Ridge Ministries and Answers in Genesis, *Global Warming*.

52. Little, "Different View on Global Warming." An Answers in Genesis employee told me that sales figures were proprietary information (personal communication, November 14, 2015). But even with those numbers, viewership would be difficult to track, for the documentary is also sold by numerous online Christian vendors, as well as Amazon.com, and is also now available online for free on several websites.

53. Rabe, "Christian Views on Global Warming"; D. James Kennedy Ministries, "Learn2Discern—United Nations Chutzpah." It is possible, if not likely, that Beisner shaped Rabe's environmental views, given that, according to Rabe's LinkedIn profile, they overlapped at Knox Theological Seminary for four years (Rabe studied for a Master of Divinity from 1999 to 2004; Beisner taught there from 2000 to 2008).

54. D. James Kennedy Ministries, "Kennedy Classics—Getting Your Priorities Straight." Fears about replacing God with government have deep historical roots. See Kruse, *One Nation under God*.

55. Ward, "Appendix B," 269.

56. Episodes airing March 21, 2008 (vol. 78), and January 25, 2010 (vol. 89). Available online at https://answersingenesis.org/media/audio/answers-with-ken-ham/ (accessed August 15, 2018).

57. Ward, "Appendix B," 259.

58. The pocket guide was advertised on the ten episodes of *Answers with Ken Ham* that aired between April 13 and April 24, 2009.

59. Oard, "Is Man the Cause of Global Warming?"

60. White, "Should We Be Concerned about Climate Change?," 198.

61. See discussion of *The Christian Worldview* below.

62. This figure is based on data pulled from SEMrush on July 28, 2017 (for the United States only). The figure is within range of AiG's own estimate of one million web visitors a month (Answers in Genesis, "Honored Again").

63. Trollinger and Trollinger, *Righting America*, 160.

64. Ham, "Presuppositions and Climate Change."

65. CNN, "Christian Right Leader"; Ward, "Appendix B," 261.

66. Williams, *God's Own Party*, 238.

67. Gilgoff, *Jesus Machine*, 61.

68. Ibid., 68.

69. Cornwall Alliance, "Open Letter."

70. The documents are "An Evangelical Declaration on Global Warming" and "Protecting the Unborn and the Pro-Life Movement from a Misleading Environmentalist Tactic."

71. FRC, "Vision and Mission Statements."

72. Saunders, "Inconvenient Ambiguity."

73. Beisner, "Cornwall Alliance Debates GW."

74. http://we-get-it.org/endorsements/ (accessed June 22, 2016; page no longer available).

75. FRC, "How a Climate Change Treaty Threatens You."

76. Perkins, "Climate Change."

77. Based on station listing as of 2016 (FRC, "Station Listing: Weekend Edition"). On weekdays at this time, *Washington Watch* aired on 182 stations (FRC, "Station Listing: Live with Tony Perkins").

78. A search conducted November 3, 2018, found twenty-one entries, the majority posted between 2007 and 2009. The posts demonstrate that the FRC was following the climate debate closely. See, for example, Perkins, "NAE's Dangerous Emissions" and "If NAE's Rich Cizik Doesn't Speak for Them."

79. Jackson and Perkins, *Personal Faith, Public Policy*, 209.

80. Ibid., 212.

81. Ibid., 211, 216.

82. Perkins, "Introduction"; FRC, *25 Pro-Family Policy Goals*, 33, 12.

83. Prentice, "Change Watch Backgrounder: Jane Lubchenco." See also FRC, "Change Watch Backgrounder: Lisa Jackson."

84. As of December 11, 2018, the kit was available online at http://downloads.frcaction.org/EF/EF10H05.pdf.

85. Inhofe was a guest on *Washington Watch* four times in 2015, each time discouraging concern about climate change (based on a search of Family Research Council podcasts on Soundcloud.com, conducted August 6, 2018).

86. *The Old-Time Gospel Hour* became *Live from Liberty* in 2004.

87. Cusic, "Television and Contemporary Christian Music," 434.

88. Williams, *God's Own Party*, 173.

89. Ibid., 178.

90. Woodruff, *Inside Politics*.

91. Falwell, "'Green' Gospel."

92. Regarding Driessen, see Cornwall Alliance, "ISA Announces Launch"; Spencer, Driessen, and Beisner, "Examination"; and Beisner et al., "Call to Truth." Regarding Wisdom, see Interfaith Stewardship Alliance, "Advisory Board."

93. Falwell, "Blood-Tainted Gift."

94. Falwell, "Evangelicals and Global Warming."

95. Estimate obtained via a SEMRush analysis conducted July 28, 2017. As of 2017, WND.com reported that it had 6.5 million unique visitors monthly (WND, "About WND").

96. Alexa, "Top Sites." Although WND is typically described as a politically conservative news aggregator, it bills itself as "the largest Christian website in the world" (WND, "About WND").

97. Ostling, "Power, Glory—and Politics." For Falwell's global warming sermons, see Media Matters for America, "Falwell Dismissed Scientific Evidence," and Allen, "Falwell Says Global Warming Tool of Satan."

98. Victory FM, "Programs"; Boodman, "How Jerry Falwell Raises His Millions"; Liberty Channel "Program Guide."

99. Sky Angel, "Sky Angel's Angel One National Network." Sandra Wagner, who was vice president for operations at the Liberty Channel at the time, confirmed that they did not track viewership during this time, and that Sky Angel declined to share its subscriber numbers. Personal communication, July 19, 2017.

100. Falwell, "'Global Warming' Fooling the Faithful"; Allen, "Falwell Says Global Warming Tool of Satan."

101. Allen, "Falwell Says Global Warming Tool of Satan."

102. Ibid.

103. Sermon notes from "The Myth of Global Warming" are available at https://web.archive.org/web/20070228175144/http://trbc.org:80/new /resources.php. The link to the guide and the guide itself can be downloaded from https://web.archive.org/web/20070228164806/http://home.trbc.org/.

104. Williams, *God's Own Party*, 270.

105. University Advancement Staff, "Ben Stein to Speak at May Commencement."

106. Mayhew, "British Advisor Speaks."

107. Ibid.

108. Liberty University, "LU Convocation FAQ."

109. See Throckmorton and Coulter, *Getting Jefferson Right*.

110. *Time*, "25 Most Influential Evangelicals in America."

111. Barton, "Science, the Bible and Global Warming."

112. WallBuilders Live, "Stations."

113. Beck, "Dangers of Environmental Extremism."

114. Barton, *Bible, Voters and the 2008 Election*, 9.

115. Grudem, *Business for the Glory of God*.

116. Grudem, "Faith and Politics."

117. Grudem, "Chapter 10c," 2.

118. Grudem, *Politics according to the Bible*, 14.

119. Ibid.

120. Clough et al., "Climate Summit."

121. This letter contrasts with some of Grudem's earlier views (e.g., the Cornwall Alliance's 2009 "Evangelical Declaration on Global Warming," which Grudem signed) in that it seems to accept the anthropogenic origins of climate change. However, it is compatible with his previous perspective in that it argues against attempting to slow the warming process (because doing so would slow economic growth and harm the poor).

122. Parshall has worked closely with the Cornwall Alliance, having signed "An Evangelical Declaration on Global Warming" (2009) and "Protect the

Poor: Ten Reasons to Oppose Harmful Climate Change Policies" (2014), served as the host of the *Resisting the Green Dragon* documentary, and hosted the launch of the Cornwall Alliance's "We Get It!" campaign. See Vu, "Christians Launch Campaign."

123. Between July 2015 and July 2016, her climate change coverage increased, with eleven full segments dedicated to climate change (Beisner was a guest on five of these programs).

124. Estimate based on a list posted on *In the Market*'s website, dated February 24, 2016. https://web.archive.org/web/20160323014000/http://www.moodyradio .org/uploadedFiles/Radio/Website_Assets/Programs/ITM-stations-list.pdf.

125. Sumser, "Conservative Talk Radio, Religious Style," 120.

126. Ibid., 107; Wrench, "Setting the Evangelical Agenda," 174.

127. Generations with Vision, "Meet Our Director." Swanson interviewed Beisner for five episodes of *Generations Radio* during the period I examined.

128. Generations Radio, "Our Story."

129. Christian Worldview, "Mission." Although Beisner has been a guest on Wheaton's show, Wheaton appears not to have signed any of the Cornwall Alliance's documents.

130. Christian Talk 660, "About Us."

131. Movieguide, "Our Story."

132. See, for example, Movieguide, "NASA Satellites Debunk Global Warming Theory"; Baehr and Snyder, "Eco-Fanatic Movies."

133. WNG Media, "2016 Media Planning Guide"; *Christianity Today*, "CT Advertising."

134. Larsen, "God's Gardeners," 318; Cornwall Alliance, "Appeal Letter" and "Board of Advisors." Olasky is also a senior fellow at the Acton Institute, the Cornwall Alliance's parent organization; see https://acton.org/about/staff/ marvin-olasky (accessed December 13, 2018).

135. Danielsen, "Fracturing over Creation Care?," 209

136. Determined via station listings on the websites for each of these programs.

137. I did ask one Southern Baptist pastor directly where he got his information; consistent with the argument I make in this chapter, he pointed me to Ken Ham.

138. Pew Research Center, "Science in America"; Jones et al., *Believers, Sympathizers and Skeptics,* 3.

139. Pew Research Center, "Science in America"; Jones et al., *Believers, Sympathizers and Skeptics,* 12. Jones and his colleagues asked respondents whether they thought climate change was a "crisis" or a "major problem."

CHAPTER EIGHT. AWAKENING THE SLEEPING GIANT

1. Djupe and Hunt identified this problem in 2009 (see Djupe and Hunt, "Beyond the Lynn White Thesis," 671). The assumption nevertheless remains widespread in the quantitative literature. See, for example, Smith and Leiserowitz, "American Evangelicals and Global Warming," or Barker and Bearce, "End-Times Theology."

2. Williams, *God's Own Party*, 153–156.

3. See chapter 7, note 9.

4. I draw especially on Kearns, "Noah's Ark," "Cooking the Truth," "Religious Climate Activism," and "Green Evangelicals"; Wilkinson, *Between God and Green;* and Bean and Teles, "Spreading the Gospel."

5. Williams, *God's Own Party*, 167.

6. Quoted in Wallis and Michaelson, "Plan to Save America," 11.

7. Balmer, *Evangelicalism*, 109–131; Williams, *God's Own Party*, 67, 117.

8. Antonio and Brulle, "Unbearable Lightness of Politics," 196; Wald and Calhoun-Brown, *Religion and Politics*, 215. The term *market fundamentalism* comes from D. Joseph Stiglitz ("Moving beyond Market Fundamentalism," 346).

9. The clash between environmentalists and the New Right was not simply incidental. The New Right's political vision extended beyond natural resource management, but Cawley showed that it had direct ties to the Sagebrush Rebellion (Cawley, *Federal Land*, 129–130). Viguerie also included the Sierra Club, the Wilderness Society, and the Environmental Defense Fund among the liberal groups that had changed the direction of the federal government for the worse (Viguerie, *New Right*, 104–105).

10. Bellant, *Coors Connection*, 84–85; Arnold, *At the Eye of the Storm*, 22.

11. Lindsay, *Faith in the Halls*, 53.

12. Arnold, *At the Eye of the Storm*, 25.

13. Viguerie, *New Right*, 57.

14. Brulle, "Institutionalizing Delay," 688.

15. Watt quotation cited in Arnold, *At the Eye of the Storm*, 158.

16. Watt, *Courage of a Conservative*, 24–25.

17. Cawley, *Federal Land*, 130–131. See also Bratton, "Ecotheology of James Watt," 230.

18. Watt, *Courage of a Conservative*, 122.

19. Ibid.

20. Ibid., 123.

21. On the growth of environmental concern within evangelical circles during this period, see Larsen, "God's Gardeners," ch. 3.

22. Schlozman, *When Movements Anchor Parties*, 80.

23. Robertson, *New World Order*, 215. The discussion in this section is indebted to Larsen, "God's Gardeners," 263–266.

24. Schlozman, *When Movements Anchor Parties*, 84; Bauer, *Our Hopes*, cited in Larsen, "God's Gardeners," 265.

25. Colson, *Dance with Deception*, 224.

26. Acton Institute, "History of Acton Institute." I use the term *Religious Right* here because Acton was founded by a Catholic, Robert Sirico.

27. Larsen, "God's Gardeners," 312. Despite his obvious debts to free-market ideology, I agree with David Gushee that Beisner is deeply motivated by theology (Gushee, *Future of Faith*, 178).

28. Bean and Teles, "Spreading the Gospel," 10. However, Catholicism's hierarchical nature does not guarantee that all parties automatically support the Pope's positions. See Agliardo, "U.S. Catholic Response to Climate Change."

29. Evangelical environmentalists' strategy for raising concern about climate change entailed moving it from the left to the center of the evangelical tradition (Wilkinson, *Between God and Green*, 25). A number of prominent evangelical environmentalists were associated with the Evangelical Left, which had long sought to challenge the Christian Right's power (Kearns, "Noah's Ark," 359; Swartz, *Moral Minority*, 3–8).

30. Hunter, *Right Wing, Wrong Bird*, 7, 10. This understanding is clear in Hunter's discussion of the ECI, in which he criticized an unnamed leader in the Christian Right (presumably James Dobson) for describing the ECI in misleading terms on a radio broadcast. He wrote that the leader's tactics were a "harmful habit [that] has crept into the way the religious right approaches points of disagreement" (79). In 2006 Hunter was named president-elect of the Christian Coalition, the Christian advocacy organization founded by Pat Robertson, but he resigned the position after its board declined to broaden its agenda to address issues like global warming (Banerjee, "Pastor Chosen to Lead").

31. *Time*, "25 Most Influential Evangelicals in America."

32. Krattenmaker, "Model of Faith," 13A.

33. Driscoll, "In a Register Interview," Special 3A.

34. Bielo, "Formed," 32.

35. Ibid., 33.

36. McLaren, *Everything Must Change*, 4.

37. Gushee, *Future of Faith*, 87, 176.

38. Krattenmaker, "Model of Faith," 13A.

39. Luo and Goodstein, "Emphasis Shifts."

40. Kirkpatrick, "Evangelical Crackup"; Avlon, "Rise"; FitzGerald, "New Evangelicals."

41. Williams, *God's Own Party*, 272; Cizik, "McCain Confronts Conservative Hurdle."

42. Colson, "Times and Evangelicals."

43. Bean and Teles, "Spreading the Gospel," 3.

44. Sider, "ESA's History."

45. Ibid.

46. Callison, *How Climate Change Comes to Matter*, 144.

47. Breslau and Brant, "God's Green Soldiers," 49; Tierney, "And on the Eighth Day," A1.

48. Larsen, "God's Gardeners," 320.

49. Antonio and Brulle, "Unbearable Lightness of Politics," 197; Dunlap and McCright, "Challenging Climate Change," 306, 300.

50. Antonio and Brulle, "Unbearable Lightness of Politics," 195.

51. Bean and Teles, "Spreading the Gospel," 3. The work of Bean and Teles on this aspect of the Christian Right's position is insightful, but I disagree with their view that the threat that action on climate change posed to the interests of other members of the Republican coalition was "in and of itself sufficient" to explain the backlash against evangelical climate change activists (ibid.). Christian Right leaders need to maintain at least some semblance of independence from the Republican Party in order to preserve their credibility among the laity. The statements of leaders in the Christian Right—including some who are very

politically entangled—demonstrates their desire to appear to be politically independent. For example, Colson wrote in 2008, "Christian leaders should not make partisan endorsements. Neither should the evangelical movement be in the hip-pocket of any political party" (Colson, "Times and Evangelicals"). Similarly, Dobson took care to appear to be apolitical even while deeply involved in politics in order to preserve his reputation among evangelicals (Williams, *God's Own Party*, 238). Leaders in the Christian Right cannot promote every issue that their coalition partners in the Republican Party champion if they are to retain their legitimacy among the laity.

52. As Kearns noted in 2007, both secular and religious climate denial activists "see environmentalists as ultimately threatening the sacred values of free enterprise and globalized capitalism" (Kearns, "Cooking the Truth," 115–116).

53. Bean and Teles, "Spreading the Gospel," 3.

54. Wilkinson, *Between God and Green*, xii. The term is a play on *creation care*.

55. Ibid., 4.

56. Pope, "Riding the Big One"; Sullivan, "Messing with God's Creation"; Goldsmith, "Editorial."

57. This recognition was what originally prompted the formation of ICES (Larsen, "God's Gardeners," 313).

58. This is not to say it did nothing. In April 2000, ICES mailed thirty-seven thousand copies of the Cornwall Declaration to potential supporters. Like later efforts to build grassroots support, this appears to have been prompted by a lack of news coverage of the release of the Cornwall Declaration (ibid., 321–322).

59. James Partnership Form 990 (Internal Revenue Service tax form), 2009. GuideStar.

60. Boston, "Inside the Values Voter Summit"; Eros and Eros, "Values Voter Summit."

61. The speech is available at www.c-span.org/video/?194449-1/values-midterm-elections (accessed July 15, 2018).

62. Although the Cornwall Alliance has played a key role in coordinating the evangelical climate skepticism campaign publicly, other organizations may have contributed to that effort. The Arlington Group, founded by Paul Weyrich and Donald Wildmon in 2003 to coordinate the Christian Right (March, "Religious Right," 6), is one likely candidate. As Peter Heltzel notes, most of those who signed the 2007 letter to the chairman of the board of the NAE criticizing Cizik were members of the Arlington Group (Heltzel, *Jesus and Justice*, 156). The Arlington Group does not disclose its membership, but in 2009 the Cornwall Alliance appointed Shannon Royce as its new executive director; according to the Cornwall Alliance's own press release about the hire, Royce had previously been the founding executive director of the Arlington Group (Cornwall Alliance, "Evangelical 'Voice of Reason'").

63. Zaleha and Szasz, "Keep Christianity Brown!," 212. See, for example, Beisner, "Are God's Resources Finite?," and Beisner, *Where Garden Meets Wilderness*.

64. Larsen, "God's Gardeners," 312. The organization was originally known as the Acton Institute for the Study of Religion and Liberty.

65. Cornwall Alliance, "Cornwall Declaration." The Cornwall Declaration drew on Beisner's book *Where the Garden Meets Wilderness,* which the Acton Institute had copublished.

66. Larsen, "God's Gardeners," 312, 318. As for funding, both the climate care and the climate backlash movements have been funded by secular organizations that aim to shape the larger public debate about climate change (see Wilkinson, *Between God and Green,* 78; and Zaleha and Szasz, "Keep Christianity Brown!," 213). Whether funding has shaped the views of each group is a matter of debate. I do not believe, as some environmentalists have alleged (see Fang, "Exclusive"), that Beisner's views were shaped by his funding sources, given his long-standing articulation of them. Jim Ball's interest in climate change certainly predates any outside funding. What seems more likely is that both groups have been able to attract outside funding because of the high stakes of the evangelical climate debate. Rather than shaping opinions, the funding's main effect has likely been to magnify the voices of the recipients.

67. Acton Institute, "Religious Leaders"; Cornwall Alliance, "Early and Notable Signers." Richard Land's name is not included on the list of early and notable signers currently available on the Cornwall Alliance's website, but he was listed as a signatory on the Acton Institute's original press release about the declaration. Larsen relates that as the Acton Institute was assembling the Cornwall Declaration, Land was working on a similar effort. He abandoned his effort and joined forces with ICES when it became clear that ICES had greater momentum (Larsen, "God's Gardeners," 319).

68. Gushee, *Future of Faith,* 177.

69. As of 2018, the publicly available tax records (Form 990s, accessed via GuideStar) for the Cornwall Alliance (filed under its parent organization, the James Partnership) showed that the Cornwall Alliance had two employees from 2014 to 2016. From 2010 to 2013, E. Calvin Beisner was the only compensated employee. The James Partnership's Form 990 for 2009 reported four employees, including Beisner (the other employees' names and roles were not described).

70. Dunlap and McCright, "Organized Climate Change Denial," 154.

71. Beisner et al., "Call to Truth"; Dunlap and McCright, "Organized Climate Change Denial," 149.

72. Ratio Christi, "Calvin Beisner Personal Profile"; Cornwall Alliance home page, accessed via the Internet Archive Wayback Machine, December 8, 2011, https://web.archive.org/web/20111208083721/http://www.cornwallalliance.org/.

73. Cornwall Alliance, "Sounding the Alarm."

74. ISA, "Letter to the National Association of Evangelicals." This was an allusion to Richard Cizik, whose contacts at the NAE had been instrumental to gaining support for the ECI. The letter succeeded in convincing Cizik to remove his name from the Call to Action (Wilkinson, *Between God and Green,* 67).

75. Goodstein, "Evangelical Leaders." The letter to the NAE is available at www.pbs.org/moyers/moyersonamerica/green/nae_response.pdf (accessed July 15, 2018).

76. The letter can be accessed via the Internet Archive Wayback Machine. See https://web.archive.org/web/20070312204056/http://www.citizenlink.org/pdfs/NAEletterfinal.pdf (accessed March 26, 2018).

77. David Harding, correspondence department manager at Family Talk, email correspondence with author, May 27, 2016.

78. ISA, "Christian Global Warming Policy News Conference"; Cornwall Alliance, "Open Letter" and "Global Warming Mainly Natural."

79. Associated Press, "In Brief," B9. The other three stories appeared in the conservative monthly *American Spectator* (Tooley, "Nothing New under the Sun"), the conservative daily *Washington Times* (Rutledge, "Evangelicals Spar over Climate," A04), and the *Kansas City Star* ("Evangelicals Question Assumptions," F13).

80. Falwell, "Evangelicals and Global Warming."

81. Cornwall Alliance, "'We Get It' Campaign Kicks Off." As of 2016 the signatories listed on www.we-get-it.org/ under the "They Get It" link were James Dobson, Wayne Grudem, Janet Parshall, Richard Land, Joel Belz, George Grant, and Jay Dennis; the singers Pat Boone and Crystal Lewis; and the politicians James Inhofe, Tom Coburn (Republican senator from Oklahoma, 2005–15), and Paul Broun (Republican representative from Georgia, 2007–15). Supporting organizations were the FRC, the Institute on Religion and Democracy, WallBuilders, the Alabama Family Alliance, the ERLC, and the Acton Institute.

82. Timing determined using the Internet Archive Wayback Machine.

83. Cornwall Alliance, "America's Leading Voice of Faith."

84. Nancy Rogers, director of donor development for the Cornwall Alliance, email communication with author, November 16, 2015. The Cornwall Alliance distributed a similar number of copies of the book associated with the DVD.

85. Wylie Carr, email communication with author, May 24, 2016. A Google alert set to identify screenings of the DVD turned up only three such events between 2013 and 2016, none of them at churches (Bernard Zaleha, email communication with author, May 23, 2016). However, since the period of examination started three years after the DVD was released, it is hard to know how significant this finding is.

86. Williams, *God's Own Party*, 16–17; National Religious Broadcasters, "History."

87. Ward, "Introduction," xviii.

88. Smith et al., *American Evangelicalism*, 34.

89. Brown and Smidt, "Media and Clergy," 80.

90. Ward, "Introduction," xviii.

91. The Evangelical Left uses primarily print and in-person speaking engagements to reach the laity (Dunn and Tyler, "Resisting Pluralism," 148).

92. Ibid.

93. Wrench, "Setting the Evangelical Agenda," 181.

94. Ward, "Introduction," xvii.

95. Stephens and Giberson, *Anointed*, 235–240.

96. Moser and Dilling, "Communicating Climate Change," 167.

97. Agyeman et al., "Climate-Justice Link."

98. Brulle et al., "Shifting Public Opinion."

99. Nisbet, "Communicating Climate Change," 15. Kearns discusses framing specifically within the context of evangelical environmentalism (Kearns, "Green Evangelicals," 164–170).

100. Stephens and Giberson, *Anointed*, 248.

101. Kahan et al., "Second National Risk and Culture Study," 5.

102. Bean and Teles, "Spreading the Gospel," 8–9.

103. Surveys show that evangelicals remained substantially more skeptical than their non-evangelical counterparts through 2016. A 2014 survey found that white evangelicals were "more likely than any other religious group to be climate change Skeptics" (Jones et al., *Believers, Sympathizers and Skeptics*, 3). A Barna poll conducted in 2016 found that only 19 percent of evangelicals believed that humans caused climate change, compared to 43 percent of practicing Christians and 53 percent of people with no faith (Barna Group, "Are Humans Responsible for Global Warming?").

EPILOGUE

1. Guth and his colleagues discuss this logic in "Faith and the Environment," 367–368.

2. For a cross-cultural comparison supporting this point, see Smith and Veldman, "Conservative Protestant Environmentalists?"

3. This theoretical point can be traced to Russell McCutcheon, who has argued that "religion" should not be understood as a self-generated category but one that was "manufactured" as a result of specific material interests (McCutcheon, *Manufacturing Religion*, xi).

4. Subramanian, "Generation Climate."

5. Egan and Mullin, "Climate Change," 216.

6. Barker and Bearce's paper and Guth and colleagues' two papers on eschatology and environmental attitudes both made this point (see Barker and Bearce, "End-Times Theology," 273; Guth et al., "Faith and the Environment," 374; and Guth et al., "Theological Perspectives," 380), but it is still insufficiently recognized by the public and in the broader literature on the social determinants of environmental attitudes.

7. Dunlap and McCright, "Climate Change Denial."

8. Brulle, Carmichael, and Jenkins, "Shifting Public Opinion."

9. Sumser, "Conservative Talk Radio," 107.

10. Providing a rough measure of this entanglement, Longo and Baker estimated that 38 percent of the effect of biblical literalism on the perceived threat of eco-catastrophe (which included climate change) was mediated by political identity. See Longo and Baker, "Economy 'versus' Environment," 355.

11. Guth et al., "Faith and the Environment," 375.

12. Guth et al., "Theological Perspectives," 380.

13. Barker and Bearce, "End-Times Theology," 270.

14. Ibid.

15. For example, according to a poll by the Pew Research Center, as of 2016, 52 percent of Republicans thought the country should do "whatever it takes to protect the environment." It is true that a much higher percentage of Democrats, 90 percent, agreed with this statement. But combining those from both parties who agree with it amounts to a large majority of the electorate—almost three-quarters of the country (Anderson, "For Earth Day").

16. Williams, *God's Own Party,* 16–17; National Religious Broadcasters, "History."

17. Ward, "Give the Winds," 115.

18. Ibid., 129; Wrench, "Setting the Evangelical Agenda," 181.

19. Viguerie, *New Right,* 120.

20. Ibid., 121–122.

21. McCright, "Political Orientation."

22. For further evidence, see Hmielowski et al., "Attack on Science?," and Feldman et al., "Climate on Cable."

23. According to Williams, it was not Falwell but Paul Weyrich who coined the term *moral majority* (Williams, *God's Own Party,* 174).

APPENDIX A. METHODS

1. Krueger and Casey, *Focus Groups,* 8, 19.

2. Morgan, Krueger, and King, *Focus Group Kit.*

3. On the other hand, a standard threat to validity in focus groups is that people can feel pressured to agree with what seems to be the dominant opinion. To guard against this, I probed for differing opinions during the discussions. I also gave participants a chance to privately write down anything they did not feel comfortable discussing in the group. This did not reveal any new opinions regarding climate change or the environment. More commonly, participants used it as an opportunity to evangelize or elaborate on views expressed during the group.

4. This designation was confirmed by checking against the list Kellstedt and Green provide in the appendix to "Knowing God's Many People."

5. Collins, "Pretesting Survey Instruments."

6. Strauss and Corbin, *Basics of Qualitative Research.*

7. I excluded volunteer organizations chiefly dedicated to litter removal, which seemed to be a civic rather than specifically environmental activity.

8. Accessed between July 1, 2016, and January 29, 2019.

Bibliography

ABC News, Time Magazine, and Stanford University. "Intensity Spikes in Concern on Warming; Many See a Change in Weather Patterns." Poll conducted by TNS Intersearch, March 9–14, 2006. https://abcnews.go.com/images/Politics/1009a1GlobalWarming.pdf.

Acton Institute. "History of Acton Institute." Grand Rapids, MI: Acton Institute. Internet Archive Wayback Machine, August 15, 2010. https://web.archive.org/web/20100815052252/www.acton.org:80/about/history-acton-institute.

———. "Religious Leaders Come Together to Promote Balanced Environmental Policies." Grand Rapids, MI: Acton Institute for the Study of Religion and Liberty, April 17, 2000. Internet Archive Wayback Machine, August 15, 2000. https://web.archive.org/web/20000815215343/www.acton.org:80/resources/comment/news.html.

Adams, Ansel. "Stop Watt! 'We Are Fighting for Our Lives.'" *Chicago Tribune,* May 26, 1981. ProQuest.

Agliardo, Michael. "The U.S. Catholic Response to Climate Change." In *How the World's Religions Are Responding to Climate Change: Social Scientific Investigations,* edited by Robin Globus Veldman, Andrew Szasz, and Randolph Haluza-DeLay, 174–192. Abingdon, UK: Routledge, 2014.

Agyeman, J., B. Doppelt, K. Lynn, and H. Hatic. "The Climate-Justice Link: Communicating Risk with Low-Income and Minority Audiences." In *Creating a Climate for Change: Communicating Climate Change and Facilitating Social Change,* edited by Susan Moser and Lisa Dilling, 119–138. Cambridge: Cambridge University Press, 2007.

Ajemian, Robert. "Zealous Lord of a Vast Domain." *Time,* March 30, 1981. Academic Search Premier.

Alexa. "The Top 500 Sites on the Web." www.alexa.com/topsites/category/Society/Religion_and_Spirituality. Accessed June 30, 2016.

———. "Top Sites in United States." www.alexa.com/topsites/countries;14/US. Accessed June 30, 2016.

Allen, Bob. "Falwell Says Global Warming Tool of Satan." Ethicsdaily.com, March 1, 2007. www.ethicsdaily.com/falwell-says-global-warming-tool-of-satan-cms-8596.

———. "How Bill Moyers Changed Baptist Journalism." *Baptist News Global*, April 20, 2015. https://baptistnews.com/article/how-bill-moyers-changed-baptist-journalism/#.Wwb_vogvyUk.

Allen, Nick. "Al Gore 'Profiting' from Climate Change Agenda." *The Telegraph*, November 3, 2009. www.telegraph.co.uk/news/earth/environment/climate change/6496196/Al-Gore-profiting-from-climate-change-agenda.html.

Alleyne, Richard. "Maybe Religion Is the Answer Claims Atheist Scientist." *The Telegraph*, September 7, 2009. www.telegraph.co.uk/journalists/richard-alleyne/6146656/Maybe-religion-is-the-answer-claims-atheist-scientist.html.

Ammerman, Nancy T. *Baptist Battles: Social Change and Religious Conflict in the Southern Baptist Convention.* New Brunswick, NJ: Rutgers University Press, 1990.

Anderson, Monica. "For Earth Day, Here's How Americans View Environmental Issues." Pew Research Center, April 20, 2017. www.pewresearch.org/fact-tank/2017/04/20/for-earth-day-heres-how-americans-view-environmental-issues/.

Answers in Genesis. "2017 Media Kit" (downloadable PDF). https://assets.answersingenesis.org/doc/articles/magazine/2017-media-kit.pdf. Accessed December 20, 2018.

———. "Answers Magazine" (webpage). https://answersingenesis.org/store/magazine/. Accessed December 20, 2018.

———. "Honored Again: AiG Website Wins Major Award." Answers in Genesis, January 6, 2012. https://answersingenesis.org/ministry-news/ministry/honored-again-aig-website-wins-major-award/.

Antonio, Robert J., and Robert J. Brulle. "The Unbearable Lightness of Politics: Climate Change Denial and Political Polarization." *The Sociological Quarterly* 52, no. 2 (2011): 195–202.

Arbuckle, Matthew. "The Interaction of Religion, Political Ideology and Concern about Climate Change in the United States." *Society and Natural Resources* 30, no. 2 (2017): 177–194.

Arbuckle, Matthew, and David Konisky. "The Role of Religion in Environmental Attitudes." *Social Science Quarterly* 96, no. 5 (November 2015): 1244–1263.

Arnold, Ron. *At the Eye of the Storm: James Watt and the Environmentalists.* Chicago: Regnery Gateway, 1982.

ASARB (Association of Statisticians of American Religious Bodies). "Southern Baptist Convention States (2010)." The ARDA. www.thearda.com/ql2010/QL_S_2010_2_1168c.asp. Accessed January 7, 2019.

Assemblies of God. "End-Time Events." Springfield, MO: General Council of the Assemblies of God. www.ag.org/top/beliefs/gendoct_17_endtime_events.cfm. Accessed October 21, 2013 (page no longer available).

———. "Our Core Doctrines: The Second Coming." Springfield, MO: General Council of the Assemblies of God. http://ag.org/top/beliefs/our_core_doctrines /second_coming/. Accessed October 21, 2013 (page no longer available).

Associated Press. "In Brief." *The Washington Post,* July 29, 2006. LexisNexis.

Avlon, John. "The Rise of the Evangelical Center." *Politico,* August 16, 2008. www.politico.com/story/2008/08/the-rise-of-the-evangelical-center-012569.

Ayers, Brandt. "Apocalypse Tomorrow." *The Cleburne News* (Heflin, AL), January 20, 2005. Access World News.

Baehr, Ted, and Tom Snyder. "Eco-Fanatic Movies Have a Mixed History at the Box Office." MovieGuide.org. www.movieguide.org/news-articles/eco-fanatic-movies-mixed-history-box-office.html. Accessed December 4, 2018.

Ball, Jim. 2003. "Ungodly Distortions." Beliefnet. www.beliefnet.com/News/2005 /03/Ungodly-Distortions.aspx. Accessed May 27, 2013.

Balmer, Randall. *Evangelicalism in America.* Waco, TX: Baylor University Press, 2016.

Banerjee, Neela. "Pastor Chosen to Lead Christian Coalition Steps Down in Dispute over Agenda." *The New York Times,* November 28, 2006. www .nytimes.com/2006/11/28/us/28pastor.html.

———. "Rev. D. James Kennedy, Broadcaster, Dies at 76." *The New York Times,* September 5, 2007. www.nytimes.com/2007/09/06/us/06kennedy.html.

———. "Southern Baptists Back a Shift on Climate Change." *The New York Times,* March 10, 2008. www.nytimes.com/2008/03/10/us/10baptist.html.

Banks, Adelle M. "Some Evangelicals Go Green, but Skepticism Lingers." *The Baptist Standard,* September 29, 2007. www.baptiststandard.com /archives/2007-archives/some-evangelicals-go-green-but-skepticism-lingers/.

Barker, David C., and David H. Bearce. "End-Times Theology, the Shadow of the Future, and Public Resistance to Addressing Global Climate Change." *Political Research Quarterly* 66, no. 2 (2012): 267–279.

Barkun, Michael. *Disaster and the Millennium.* New Haven, CT: Yale University Press, 1974.

———. "Divided Apocalypse: Thinking about the End in Contemporary America." *Soundings: An Interdisciplinary Journal* 66, no. 3 (1983): 257–280.

Barna Group. "Are Humans Responsible for Global Warming?" Barna Group, September 22, 2016. www.barna.com/research/humans-responsible-global-warming/.

Barton, David. *The Bible, Voters and the 2008 Election.* Aledo, TX: WallBuilders, 2008. www.wallbuilders.com/wp-content/uploads/2017/02/BibleVoters_lowres.pdf.

———. "Science, the Bible and Global Warming." Undated MP3 recording, 65:02. Aledo, TX: WallBuilders. https://shop.wallbuilders.com/science-the-bible-global-warming-mp3.

Bauer, Gary. *Our Hopes, Our Dreams: A Vision for America.* Colorado Springs, CO: Focus on the Family, 1996.

Beal, R. S., Jr. "Can a Premillennialist Consistently Entertain a Concern for the Environment? A Rejoinder to Al Truesdale." *Perspectives on Science and Christian Faith* 46, no. 3 (1994): 172–177.

Bean, Lydia. *The Politics of Evangelical Identity: Local Churches and Partisan Divides in the United States and Canada.* Princeton, NJ: Princeton University Press, 2014.

Bean, Lydia, and Steve Teles. "Spreading the Gospel of Climate Change: An Evangelical Battleground." Washington, DC: New America, 2015. https://static.newamerica.org/attachments/11649-spreading-the-gospel-of-climate-change/climate_care11.9.4f0142a50aa24a2ba65020f7929f6fd7.pdf.

Bebbington, David W. *Evangelicalism in Modern Britain: A History from the 1730s to the 1980s.* London: Routledge, 1989.

Beck, Glenn. "Dangers of Environmental Extremism." FoxNews.com, October 15, 2010. Rush transcript. www.foxnews.com/story/2010/10/18/glenn-beck-dangers-environmental-extremism.html.

Begley, Sharon, Eve Conant, Sam Stein, Eleanor Clift, and Matthew Philips. 2007. "The Truth about Denial." *Newsweek*, August 13, 2007, 20–29.

Beisner, E. Calvin. "Are God's Resources Finite? A Group of Christian Leaders Claim They Are, but Does the Claim Square with the Evidence?" *World* 8, no. 27 (1993): 8–10.

———. "Cornwall Alliance Debates GW at Family Research Council." *Acton Institute Powerblog*, June 1, 2007. http://blog.acton.org/archives/1685-cornwall-alliance-debates-gw-at-family-research-council.html.

———. *Where Garden Meets Wilderness: Evangelical Entry into the Environmental Debate.* Grand Rapids, MI: Acton Institute for the Study of Religion and Liberty and W. B. Eerdmans, 1997.

Beisner, E. Calvin, Paul K. Driessen, Ross McKitrick, and Roy W. Spencer. "A Call to Truth, Prudence, and Protection of the Poor: An Evangelical Response to Global Warming." Burke, VA: Cornwall Alliance for the Stewardship of Creation. July 25, 2006. www.cornwallalliance.org/docs/a-call-to-truth-prudence-and-protection-of-the-poor.pdf.

Bellant, Russ. *The Coors Connection: How Coors Family Philanthropy Undermines Democratic Pluralism.* Brooklyn, NY: South End Books, 1991.

Bergmann, Sigurd. "Climate Change Changes Religion: Space, Spirit, Ritual, Technology—through a Theological Lens." *Studia Theologica* 63, no. 2 (October 2009): 98–118.

Berry, Evan. *Devoted to Nature: The Religious Roots of American Environmentalism.* Oakland: University of California Press, 2015.

Bielo, James. "Formed: Emerging Evangelicals Navigate Two Transformations." In *The New Evangelical Social Engagement*, edited by Brian Steensland and Philip Goff, 31–49. New York: Oxford University Press, 2014.

Boodman, Sandra G. "How Jerry Falwell Raises His Millions: The Fundraising Technique." *The Washington Post*, June 28, 1981. www.washingtonpost.com/archive/politics/1981/06/28/how-falwell-raises-his-millions-the-fundraising-technique/658e20c4-5298-42c7-bc01-1f80fdf21285/?utm_term=.b184c61a267b.

Boston, Rob. "Inside the Values Voter Summit." *Church and State*, November 2006. www.au.org/church-state/november-2006-church-state/featured/inside-the-values-voter-summit.

Botelho, Greg. "Weeks-Long Wildfire Forces Closure of Vast Okefenokee Swamp." CNN, July 3, 2011. www.cnn.com/2011/US/07/02/georgia.swamp .fire/index.html.

Boyd, Heather Hartwig. "Christianity and the Environment in the American Public." *Journal for the Scientific Study of Religion* 38, no. 1 (1999): 36–44.

Boyer, Paul. *When Time Shall Be No More: Prophecy Belief in Modern American Culture.* Cambridge, MA: Harvard University Press, 1992.

Bratton, Susan Power. "The Ecotheology of James Watt." *Environmental Ethics* 5 (Fall 1983): 225–236.

BreakPoint. "About." www.breakpoint.org/about/. Accessed August 3, 2018.

Breslau, Karen, and Martha Brant. "God's Green Soldiers." *Newsweek,* February 13, 2006. Access World News.

Brinkley, Alan. "Clear and Present Dangers." Review of *American Theocracy: The Peril and Politics of Radical Religion, Oil, and Borrowed Money in the 21st Century,* by Kevin Phillips. *The New York Times,* March 19, 2006. www.nytimes.com/2006/03/19/books/review/clear-and-present-dangers .html.

Brinton, Henry G. "Green, Meet God." *USA Today,* November 10, 2008. LexisNexis.

Brown, Donald W., and Corwin E. Smidt. "Media and Clergy: Influencing the Influential?" *Journal of Media and Religion* 2, no. 2003 (2003): 75–92.

Brulle, Robert J. "Institutionalizing Delay: Foundation Funding and the Creation of U.S. Climate Change Counter-Movement Organizations." *Climatic Change,* no. 122 (2014): 681–694.

Brulle, Robert J., Jason Carmichael, and J. Craig Jenkins. "Shifting Public Opinion on Climate Change: An Empirical Assessment of Factors Influencing Concern over Climate Change in the U.S., 2002–2010." *Climatic Change* 114, no. 2 (2012): 169–188.

Buettner, Dan. *The Blue Zones: 9 Lessons for Living Longer from the People Who've Lived the Longest,* 2nd ed. Washington, DC: National Geographic, 2012.

Callicott, J. Baird. "A Brief History of Conservation Philosophy." In *Sustainable Ecological Systems: Implementing an Ecological Approach to Land Management,* coordinated by Wallace Covington and Leonard DeBano. Fort Collins, CO: USDA, Forest Service, and Rocky Mountain Forest and Range Experiment Station, General Technical Report RM-247, 1993, 10–14. www .fs.fed.us/rm/pubs_rm/rm_gtr247/rm_gtr247_010_014.pdf.

Callison, Candis. *How Climate Change Comes to Matter: The Communal Life of Facts.* Durham, NC: Duke University Press, 2014.

Campolo, Anthony. *How to Rescue the Earth without Worshipping Nature.* Nashville, TN: Thomas Nelson, 1992.

Carpenter, Joel A. "Fundamentalist Institutions and the Rise of Evangelical Protestantism, 1929–1942." *Church History* 49, no. 1 (1980): 62–75.

Carr, Wylie. "The Faithful Skeptics: Conservative Christian Religious Beliefs and Perceptions of Climate Change." Master's thesis, University of Montana, 2010.

Carr, Wylie, Michael Patterson, Laurie Yung, and Daniel Spencer. "The Faithful Skeptics: Evangelical Religious Beliefs and Perceptions of Climate Change." *Journal for the Study of Religion, Nature and Culture* 6, no. 3 (November 2012): 276–299.

Castelli, Elizabeth A. "Persecution Complexes: Identity Politics and the 'War on Christians.'" *Differences: A Journal of Feminist Cultural Studies* 18, no. 3 (2007): 152–180.

Castelli, Jim. "The Environmental Gospel according to James Watt." *Chicago Tribune*, October 25, 1981. ProQuest.

———. "What's Watt with God and the Interior?" *The Washington Star*, July 25, 1981, A8. DC Public Library.

Cawley, R. McGreggor. *Federal Land, Western Anger: The Sagebrush Rebellion and Environmental Politics*. Lawrence: University of Kansas Press, 1993.

CBN (Christian Broadcasting Network). "Advertise with CBN." https://secure .cbn.com/contact/feedback-advertise-with-cbn.aspx?mobile=false. Accessed July 28, 2017.

———. "Al Gore—January 31, 1992." January 31, 1992. www1.cbn.com /content/al-gore-%E2%80%93-january-31-1992.

———. "CBN NewsWatch—December 4, 2009." December 4, 2009. www1 .cbn.com/content/cbn-newswatch-december-4-2009 (page no longer available).

CBN News. "Team Effort to Save the Planet." March 30, 2008. www.cbn.com /cbnnews/us/2008/march/team-effort-to-save-the-planet-/?mobile=false. Accessed May 30, 2016 (page no longer available).

Christianity Today. "CT Advertising." www.christianitytodayads.com /christianity-today/. Accessed December 4, 2018.

Christian Post. "The *Christian Post* Media Kit." [2013?]. www.christianpost .com/services/christian-advertising/downloads/Mediakit_2013.pdf. Accessed July 28, 2017.

Christian Talk 660. "About Us." www.christiantalk660.com/about/us/. Accessed December 4, 2018.

Christian Worldview. "Mission." Internet Archive Wayback Machine, March 7, 2016. https://web.archive.org/web/20160307204912/www.thechristian worldview.org/test.

Churches of Christ. "Internet Ministries" (home page). www.church-of-christ .org/. Accessed May 17, 2017.

Cizik, Richard. "McCain Confronts Conservative Hurdle." Interview by Liane Hansen. *Weekend Edition*, National Public Radio, February 10, 2008. www .npr.org/templates/story/story.php?storyId=18854833.

———. "Natural Disasters, Climate Change, and the 'Burn It All Downers.'" *The Washington Post*, May 21, 2013. ProQuest.

Clarkson, Frederick. "Christian Reconstructionists and Dominionists." In *Encyclopedia of Millennialism and Millennial Movements*, edited by Richard A. Landes, 84–88. New York: Routledge, 2000.

Clough, Charles, Neil Frank, Wayne Grudem, Jeffrey Haymond, Tracy Miller, Roy Spencer, and David Wells. "Climate Summit: Why We Believe Paris Proposals Doom Billions to Live in Extreme Poverty." FoxNews.com, December 7, 2015. www.foxnews.com/opinion/2015/12/07/climate-summit-why-believe-paris-proposals-doom-billions-to-live-in-extreme-poverty.html.

Clouse, Robert G., Robert N. Hosack, and Richard V. Pierard. *The New Millennium Manual: A Once and Future Guide*. Grand Rapids, MI: BridgePoint Books, 1999.

CNN. "Christian Right Leader Writes off Giuliani." May 18, 2007. www.cnn.com/2007/POLITICS/05/17/giuliani.dobson/index.html?_s=PM:POLITICS.

———. "Southern Baptist Leaders Shift Position on Climate Change." March 10, 2008. www.cnn.com/2008/US/03/10/baptist.climate/.

Cohen, Richard, and Peter Bell. "Congressional Insiders Poll." *National Journal* 39, no. 5 (February 2007): 6–7.

Collins, Debbie. "Pretesting Survey Instruments: An Overview of Cognitive Methods." *Quality of Life Research* 12, no. 3 (2003): 229–238.

Colson, Chuck. "3 Reasons the Copenhagen Climate Summit Is Lacking." Colson Center for a Christian Worldview, December 1, 2009. www.colsoncenter.org/twominutewarning/archive/entry/33/15942. Accessed May 24, 2016 (page no longer available; also found at www.youtube.com/watch?v=w_tI7MDbjm8).

———. "BreakPoint: Global Warming and Ideology." October 26, 2010. http://www.breakpoint.org/2010/10/breakpoint-global-warming-ideology/ (originally accessed June 30, 2016, at http://www.breakpoint.org/bpcommentaries/entry/13/15626; page no longer available).

———. "BreakPoint: Global Warming as Religion." December 2, 2009. www.breakpoint.org/2009/12/breakpoint-global-warming-religion/ (originally accessed May 24, 2016, at www.breakpoint.org/bpcommentaries/entry/13/13898; page no longer available).

———. "BreakPoint: 'Not Evil Just Wrong.'" October 15, 2009. www.breakpoint.org/2009/10/breakpoint-not-evil-just-wrong/.

———. "Engineering the Earth." *The Christian Post*, October 25, 2010. www.christianpost.com/news/engineering-the-earth.html.

———. "Global Warming as Religion." *The Christian Post*, December 4, 2009. www.christianpost.com/news/global-warming-as-religion.html.

———. "The Times and Evangelicals." *The Christian Post*, January 16, 2008. www.christianpost.com/news/the-times-and-evangelicals-30879/.

Colson, Chuck, and Nancy R. Pearcey. *A Dance with Deception: Revealing the Truth behind the Headlines*. Nashville, TN: Thomas Nelson, 1993.

Committee on Interior and Insular Affairs. *Reorganization of the Office of Surface Mining: Oversight Hearing before the Subcommittee on Energy and the Environment* (serial no. 97–13). July 16, 1981. Washington, DC: U.S. Government Printing Office. HeinOnline.

Connable, Sean. "The 'Christian Nation' Thesis and the Evangelical Echo Chamber." In *The Electronic Church in the Digital Age: Cultural Impacts of Evangelical Mass Media*, vol. 2, edited by Mark Ward Sr., 183–203. Santa Barbara, CA: ABC-CLIO, 2015.

Cook, John, Naomi Oreskes, Peter Doran, William Anderegg, Bart Verheggen, Ed Maibach, J. Stuart Carlton, Stephan Lewandowsky, Andrew Skuce, Sarah Green, Dana Nuccitelli, Peter Jacobs, Mark Richardson, Bärbel

Winkler, Rob Painting, and Ken Rice. "Consensus on Consensus: A Synthesis of Consensus Estimates on Human-Caused Global Warming." *Environmental Research Letters* 11, no. 4 (2016): 1–7.

Coral Ridge Ministries and Answers in Genesis. *Global Warming: A Scientific and Biblical Exposé of Climate Change.* Produced by Ben Wilt, Jason Beaupied, and Javier Peña, 2008. Video, 50:00.

Corbin, Juliet, and Anselm Strauss. "Grounded Theory Research: Procedures, Canons, and Evaluative Criteria." *Qualitative Sociology* 13, no. 1 (1990): 3–21.

Cornwall Alliance for the Stewardship of Creation. "America's Leading Voice of Faith on Stewardship Issues Announces New Initiative." Press release, April 20, 2010. Burke, VA: Cornwall Alliance. https://web.archive.org/web/20100626140847/www.cornwallalliance.org/docs/americas-leading-voice-of-faith-on-stewardship-issues-announces-new-initiative-to-expose-serious-dangers-of-green-dragon-environmentalism.pdf.

———. "Appeal Letter to the National Association of Evangelicals on the Issue of Global Warming." January 31, 2006. Burke, VA: Cornwall Alliance. http://cornwallalliance.org/2006/01/appeal-letter-to-the-national-association-of-evangelicals-on-the-issue-of-global-warming/.

———. "Board of Advisors." Burke, VA: Cornwall Alliance. Internet Archive Wayback Machine, February 9, 2008. https://web.archive.org/web/20080209135816/http:/www.cornwallalliance.org/about/board-of-advisors/.

———. "The Cornwall Declaration on Environmental Stewardship." Burke, VA: Cornwall Alliance. https://cornwallalliance.org/landmark-documents/the-cornwall-declaration-on-environmental-stewardship/. Accessed April 12, 2018.

———. "Early and Notable Signers of the Cornwall Declaration." October 29, 1999. Burke, VA: Cornwall Alliance. https://cornwallalliance.org/1999/10/notable-signers-of-the-cornwall-declaration/.

———. "Evangelical Declaration on Global Warming." May 1, 2009. Burke, VA: Cornwall Alliance. https://cornwallalliance.org/2009/05/evangelical-declaration-on-global-warming/.

———. "Evangelical 'Voice of Reason' on Climate Change Staffs Up with Addition of Top-Flight Executive Director to Expand National Outreach." Press release, April 22, 2009. Burke, VA: Cornwall Alliance. Internet Archive Wayback Machine, November 16, 2009. https://web.archive.org/web/20091116125741/www.cornwallalliance.org:80/press/read/evangelical-voice-of-reason-on-climate-change-staffs-up-with-addition-of-top-flight-executive-director-to-expand-national-outreach/.

———. "Global Warming Mainly Natural and Not Catastrophic, Says New Study from Interfaith Stewardship Alliance." Press release, August 2, 2006. Burke, VA: Cornwall Alliance. www.archive.cornwallalliance.org/2006/08/02/global-warming-mainly-natural-and-not-catastrophic-says-new-study-from-interfaith-stewardship-alliance/.

———. "ISA Announces Launch of Cornwall Network at Senate Luncheon—Driessen." April 19, 2006. Burke, VA: Cornwall Alliance. www.archive.cornwallalliance.org/2006/04/19/isa-announces-launch-of-cornwall-network-at-senate-luncheon-driessen/.

———. "An Open Letter to the Signers of 'Climate Change: An Evangelical Call to Action' and Others Concerned about Global Warming." July 2006. Burke, VA: Cornwall Alliance. www.cornwallalliance.org/docs/an-open-letter-to-the-signers-of-climate-change-an-evangelical-call-to-action-and-others-concerned-about-global-warming.pdf.

———. "Protecting the Unborn and the Pro-Life Movement from a Misleading Environmentalist Tactic." February 8, 2012. Burke, VA: Cornwall Alliance. www.cornwallalliance.org/2012/02/protecting-the-unborn-and-the-pro-life-movement-from-a-misleading-environmentalist-tactic-2/.

———. "Protect the Poor: Ten Reasons to Oppose Harmful Climate Change Policies." Accessed April 16, 2019. Burke, VA: Cornwall Alliance. www.cornwallalliance.org/landmark-documents/protect-the-poor-ten-reasons-to-oppose-harmful-climate-change-policies/.

———. "Sounding the Alarm about Dangerous Environmental Extremism." Press release, November 10, 2010. Burke, VA: Cornwall Alliance. Internet Archive Wayback Machine. https://web.archive.org/web/20110122094806/www.cornwallalliance.org:80/press/read/sounding-the-alarm-about-dangerous-environmental-extremism/.

———. "'We Get It' Campaign Kicks Off Effort to Enlist One Million Signers on Landmark Statement." Press release, May 15, 2008. Burke, VA: Cornwall Alliance. Internet Archive Wayback Machine, May 17, 2008. https://web.archive.org/web/20080517090618/http:/www.cornwallalliance.org:80/alert/we-get-it-campaign-kicks-off-effort-to-enlist-one-million-signers-on-landmark-statement/.

Coulter, Ann. *Godless: The Church of Liberalism*. New York: Random House, 2006.

Crouch, Andy. "Environmental Wager: Why Evangelicals Are—but Shouldn't Be—Cool toward Global Warming." *Christianity Today* 49, no. 8 (August 2005): 66. www.christianitytoday.com/ct/2005/august/22.66.html.

Curry, Janel. "Christians and Climate Change: A Social Framework Analysis." *Perspectives on Science and Christian Faith* 60, no. 3 (2008): 156–164.

Curry-Roper, Janel M. "Contemporary Christian Eschatologies and Their Relation to Environmental Stewardship." *Professional Geographer* 42, no. 2 (1990): 157–169.

Cusic, Don. "Television and Contemporary Christian Music." In *Encyclopedia of Contemporary Christian Music: Pop, Rock and Worship,* edited by Don Cusic, 432–437. Santa Barbara, CA: ABC-CLIO, 2010.

Danielsen, Sabrina. "Fracturing over Creation Care? Shifting Environmental Beliefs among Evangelicals, 1984–2010." *Journal for the Scientific Study of Religion* 52, no. 1 (2013): 198–215.

Davenport, Coral. "As Climate Risks Rise, Talks in Paris Set Stage for Action." *The New York Times,* November 30, 2015. www.nytimes.com/2015/11/30/us/politics/paris-climate-talks.html.

Davidson, Osha Gray. "Dirty Secrets." *Mother Jones* 28, no. 5 (September–October 2003): 48–53.

Delfattore, Joan. *The Fourth R: Conflicts over Religion in America's Public Schools.* New Haven, CT: Yale University Press, 2004.

Dewar, Helen, and Richard L. Lyons. "All but One of President's Department Heads Confirmed by Senate." *The Washington Post,* January 23, 1981. ProQuest.

DeWitt, Calvin. "Creation's Environmental Challenge to Evangelical Christianity." In *The Care of Creation: Focusing Concern and Action,* edited by R. J. Berry, 60–73. Leicester, UK: Inter-Varsity Press, 2000.

D. James Kennedy Ministries. "Kennedy Classics—Getting Your Priorities Straight." Kennedy Classics, April 25, 2014. Video, 28:30. www.youtube.com /watch?v=jFtDjSHz1DY.

———. "Learn2Discern—United Nations Chutzpah." January 8, 2010. Video, 01:30. www.youtube.com/watch?v=_FMO2owu8UY.

Djupe, Paul A., and Patrick K. Hunt. "Beyond the Lynn White Thesis: Congregational Effects on Environmental Concern." *Journal for the Scientific Study of Religion* 48, no. 4 (2009): 670–686.

Dolan, Eric W. "Belief in Biblical End-Times Stifling Climate Change Action in U.S.: Study." *Raw Story,* May 1, 2013. www.rawstory.com/2013/05/belief-in-end-times-stifling-climate-change-action-in-u-s-study/.

Driscoll, Gwendolyn. "In a Register Interview, Saddleback's Pastor Talks of PEACE, Politics and a New 'Purpose-Driven' Book." *Orange County Register,* December 31, 2006. Access World News.

Driscoll, Mark. "Catalyst, Comedy and Critics." *Pastor Mark Driscoll* (blog), May 15, 2013. Internet Archive Wayback Machine. https://web.archive.org /web/20130608111917/pastormark.tv/2013/05/15/catalyst-comedy-and-critics.

Dryzek, John S. *The Politics of the Earth: Environmental Discourses.* Oxford: Oxford University Press, 1997.

Dunlap, Riley E., and Aaron M. McCright. "Challenging Climate Change: The Denial Countermovement." In *Climate Change and Society: Sociological Perspectives,* edited by Riley Dunlap and Robert Brulle, 300–332. New York: Oxford University Press, 2015.

———. "Climate Change Denial: Sources, Actors and Strategies." In *Routledge Handbook of Climate Change and Society,* edited by Constance Lever-Tracy, 240–259. Abingdon, UK: Routledge, 2010.

———. "Organized Climate Change Denial." In *The Oxford Handbook of Climate Change and Society,* edited by John S. Dryzek and Richard B. Norgaard, 144–160. Oxford: Oxford University Press, 2011.

———. "A Widening Gap: Republican and Democratic Views on Climate Change." *Environment: Science and Policy for Sustainable Development* 50, no. 5 (2008): 26–36.

Dunlap, Riley E., Kent D. Van Liere, Angela G. Mertig, and Robert Emmet Jones. "Measuring Endorsement of the New Ecological Paradigm: A Revised NEP Scale." *Journal of Social Issues* 56 no. 3 (2000): 425–442.

Dunn, Scott W., and J. Adam Tyler. "Resisting Pluralism: Evangelical Media Framing of the Role of Faith in American Politics." In *The Electronic Church in the Digital Age: Cultural Impacts of Evangelical Mass Media,* vol. 2, edited by Mark Ward Sr., 131–154. Santa Barbara, CA: ABC-CLIO, 2016.

Eagle Forum. "Global Warming." https://eagleforum.org/topics/global-warming/. Accessed December 12, 2018.

ECI (Evangelical Climate Initiative). "Climate Change: An Evangelical Call to Action." February 8, 2006. www.npr.org/documents/2006/feb/evangelical /calltoaction.pdf.

Eckberg, Douglas L., and T. Jean Blocker. "Christianity, Environmentalism, and the Theoretical Problem of Fundamentalism." *Journal for the Scientific Study of Religion* 35, no. 4 (1996): 343–355.

Ecklund, Elaine Howard, Chrisopher P. Scheitle, Jared Peifer, and Daniel Bolger. "Examining Links between Religion, Evolution Views, and Climate Change Skepticism." *Environment and Behavior* 49, no. 9 (2017): 985–1006.

Edwards, David. "Limbaugh: Christians 'Cannot Believe in Manmade Global Warming.'" *Raw Story*, August 12, 2013. www.rawstory.com/2013/08 /limbaugh-christians-cannot-believe-in-manmade-global-warming.

EEN (Evangelical Environmental Network). "Our Witness." www.creationcare .org/witness. Accessed December 27, 2018.

Egan, Patrick, and Megan Mullin. "Climate Change: US Public Opinion." *Annual Review of Political Science* 20 (May 2017): 209–227.

Ellingson, Stephen. *To Care for Creation: The Emergence of the Religious Environmental Movement*. Chicago: University of Chicago Press, 2016.

Ellison Research. "Nationwide Survey Shows Concerns of Evangelical Christians over Global Warming." Prepared for the Evangelical Environmental Network, February 8, 2006. www.npr.org/documents/2006/feb/evangelical /newsrelease.pdf.

Elsasser, Shaun W., and Riley E. Dunlap. "Leading Voices in the Denier Choir: Conservative Columnists' Dismissal of Global Warming and Denigration of Climate Science." *American Behavioral Scientist* 57, no. 6 (2013): 754–776.

Emerson, Michael O., and Christian Smith. *Divided by Faith: Evangelical Religion and the Problem of Race in America*. Oxford: Oxford University Press, 2000.

ERLC Staff. "Calling Southern Baptists to Be 'Salt' and 'Light.'" *SBC Life: Journal of the Southern Baptist Convention*, December 2011. www.sbclife .net/Articles/2011/12/sla6.

Eros, Sherry, and Steve Eros. "Values Voter Summit Takes on Liberals." *Human Events* 62, no. 33 (September 29, 2006). LexisNexis.

Ertelt, Steven. "Super Bowl Ad Gets Focus on the Family Big Traffic for Tebow Non-abortion Story." LifeNews.com, February 9, 2010. www.lifenews .com/2010/02/09/nat-5981/.

Eskridge, Larry. "Defining the Term in Contemporary Times." Wheaton, IL: Wheaton College Institute for the Study of American Evangelicals, 2014. www.wheaton.edu/ISAE/Defining-Evangelicalism/Defining-the-Term (page no longer available).

Faith and Action in the Nation's Capital. "Senior Senator James Inhofe Talks about His Unusual Love for Africa." July 25, 2009. www.faithandaction .org/senior-senator-james-inhofe-talks-about-his-unusual-love-for-africa/ (page no longer available).

Falwell, Jerry. "A Blood-Tainted Gift to Evangelicals." WND, March 4, 2006. www.wnd.com/2006/03/35086/.

———. "Evangelicals and Global Warming." WND, November 18, 2006. www.wnd.com/2006/11/38937/.

———. "'Global Warming' Fooling the Faithful." WND, May 26, 2007. www.wnd.com/2007/02/40332/.

———. "'Green' Gospel." WND, February 11, 2006. www.wnd.com/2006/02/34769.

Fang, Lee. "Exclusive: The Oily Operators behind the Religious Climate Change Denial Front Group, Cornwall Alliance." ThinkProgress, June 15, 2010. https://thinkprogress.org/exclusive-the-oily-operators-behind-the-religious-climate-change-denial-front-group-cornwall-6caf65708c53/.

Feldman, Lauren, Edward W. Maibach, Connie Roser-Renouf, and Anthony Leiserowitz. "Climate on Cable: The Nature and Impact of Global Warming Coverage on Fox News, CNN, and MSNBC." International Journal of Press and Politics 17, no. 1 (2012): 3–31.

Fine, Gary. "Notorious Support: The America First Committee and the Personalization of Policy." Mobilization: An International Quarterly 11, no. 4 (2006): 405–426.

FitzGerald, Frances. "The New Evangelicals." The New Yorker, June 30, 2008. www.newyorker.com/magazine/2008/06/30/the-new-evangelicals.

Foust, Michael. "SBC Leaders: Caring for Earth a Command." Baptist Press, March 19, 2008. www.bpnews.net/27659/sbc-leaders-caring-for-earth-a-command.

Fowler, Robert Booth. The Greening of Protestant Thought. Chapel Hill: University of North Carolina Press, 1995.

Fox, Stephen. John Muir and His Legacy: The American Conservation Movement. Boston: Little, Brown, 1981.

FRC (Family Research Council). 25 Pro-Family Policy Goals for the Nation, edited by Peter Sprigg. Washington, DC: Family Research Council, 2008. https://downloads.frc.org/EF/EF11D02.pdf.

———. "Change Watch Backgrounder: Lisa Jackson." February 10, 2009. www.frcblog.com/2009/02/change-watch-backgrounder-lisa-jackson/.

———. "How a Climate Change Treaty Threatens You, Your Nation, and Your Church." Internet Archive Wayback Machine, April 22, 2010. https://web.archive.org/web/20100422055325/www.frc.org/eventregistration/how-a-climate-change-treaty-threatens-you-your-nation-and-your-church.

———. "Station Listing: Live with Tony Perkins." Internet Archive Wayback Machine, January 4, 2016. https://web.archive.org/web/20160104231413/www.frc.org/live-with-tony-perkins-station-listings.

———. "Station Listing: Weekend Edition." Internet Archive Wayback Machine, January 26, 2016. https://web.archive.org/web/20160126041802/www.frc.org/weekend-edition-station-listings.

———. "Vision and Mission Statements." www.frc.org/mission-statement. Accessed December 22, 2018.

Funk, Cary, and Becka A. Alper. "Religion and Views on Climate and Energy Issues." Pew Research Center, October 22, 2015. www.pewinternet.org/2015/10/22/religion-and-views-on-climate-and-energy-issues/.

Gauchat, Gordon. "Politicization of Science in the Public Sphere: A Study of Public Trust in the United States, 1974–2010." *American Sociological Review* 77, no. 2 (2012): 167–187.

Generations Radio. "Our Story." http://generationsradio.com/. Accessed December 4, 2018.

Generations with Vision. "Meet Our Director." Internet Archive Wayback Machine, June 13, 2015. https://web.archive.org/web/20150603011745 /https://generationswithvision.com/about/meet-our-director/.

Gilgoff, Dan. *The Jesus Machine: How James Dobson, Focus on the Family and Evangelical America Are Winning the Culture War.* New York: St. Martin's Press, 2008.

Glacken, Clarence. *Traces on the Rhodian Shore: Nature and Culture in Western Thought from Ancient Times to the End of the Eighteenth Century.* Berkeley: University of California Press, 1967.

Glaser, Barney, and Anselm Strauss. *The Discovery of Grounded Theory: Strategies for Qualitative Research.* New York: Routledge, 2017 [1967].

Goldenberg, Suzanne. "The Worst of Times: Bush's Environmental Legacy Examined." *The Guardian,* January 16, 2009. www.theguardian.com/politics /2009/jan/16/greenpolitics-georgebush.

Goldsmith, Zac. "Editorial." *The Ecologist* 36, no. 2 (April 2006): 5. EBSCO.

Goodstein, Laurie. "Evangelical Leaders Join Global Warming Initiative." *The New York Times,* February 8, 2006. www.nytimes.com/2006/02/08/us /evangelical-leaders-joinglobal-warming-initiative.html.

Gore, Al. *Earth in the Balance: Ecology and the Human Spirit.* Boston: Houghton Mifflin, 1992.

Greeley, Andrew. "Religion and Attitudes toward the Environment." *Journal for the Scientific Study of Religion* 32, no. 1 (1993): 19–28.

Greeley, Andrew, and Michael Hout. *The Truth about Conservative Christians: What They Think and What They Believe.* Chicago: University of Chicago Press, 2006.

Green, John C. "The American Religious Landscape and Political Attitudes: A Baseline for 2004." Pew Forum on Religion and Public Life, Washington, DC, 2004. www.uakron.edu/bliss/research/archives/2004/Religious_Landscape_ 2004.pdf (page no longer available).

———. "The Fifth National Survey on Religion and Politics: A Baseline for the 2008 Presidential Election." Akron, OH: Ray C. Bliss Institute of Applied Politics, 2008. www.uakron.edu/bliss/research/archives/2008/Fifth_National_ Survey_Religion_Politics.pdf (page no longer available).

Griffith, R. Marie. *God's Daughters: Evangelical Women and the Power of Submission.* Berkeley: University of California Press, 1997.

Grudem, Wayne. *Business for the Glory of God: The Bible's Teaching on the Moral Goodness of Business.* Wheaton, IL: Crossway Books, 2003.

———. "Chapter 10c—The Environment: Global Warming." Outline of sermon delivered November 21, 2010. www.christianessentialssbc.com/down loads/2010/2010-11-21.pdf.

———. "Faith and Politics: Government and Economics—Global Warming." Sermon delivered November 21, 2010. www.christianessentialssbc.com /messages/.

———. *Politics according to the Bible: A Comprehensive Resource for Understanding Modern Political Issues according to Scripture*. Grand Rapids, MI: Zondervan, 2010.

———. *Systematic Theology*. Leicester, UK: Inter-Varsity Press and Zondervan, 1994.

Gushee, David. *The Future of Faith in American Politics: The Public Witness of the Evangelical Center*. Waco, TX: Baylor University Press, 2008.

Guth, James, John C. Green, Lyman A. Kellstedt, and Corwin E. Smidt. "Faith and the Environment: Religious Beliefs and Attitudes on Environmental Policy." *American Journal of Political Science* 39, no. 2 (May 1995): 364–382.

Guth, James, Lyman A. Kellstedt, Corwin E. Smidt, and John Green. "Theological Perspectives and Environmentalism among Religious Activists." *Journal for the Scientific Study of Religion* 32, no. 4 (December 1993): 373–382.

Haberman, David. *River of Love in an Age of Pollution: The Yamuna River of Northern India*. Berkeley: University of California Press, 2006.

Hagerty, Barbara B. "Climate Change Prompts Debate among Baptists." National Public Radio, March 10, 2008. www.npr.org/templates/story /story.php?storyId=88074916.

Haluza-DeLay, Randolph. "Churches Engaging the Environment: An Autoethnography of Obstacles and Opportunities." *Human Ecology Review* 15, no. 1 (2008): 71–81.

Ham, Ken. "Presuppositions and Climate Change." *Answers with Ken Ham* (radio transcript), May 1, 2017. https://answersingenesis.org/media/audio /answers-with-ken-ham/volume-125/presuppositions-and-climate-change/.

Hand, Carl M., and Kent D. van Liere. "Religion, Mastery-over-Nature, and Environmental Concern." *Social Forces* 63, no. 2 (1984): 555–570.

Harden, Blaine. "The Greening of Evangelicals: The Christian Right Turns, Sometimes Warily, to Environmentalism." *The Washington Post*, February 6, 2005. www.washingtonpost.com/wp-dyn/articles/A1491–2005Feb5.html.

Harding, Susan. "Representing Fundamentalism: The Problem of the Repugnant Cultural Other." *Social Research* 58, no. 2 (1991): 373–393.

Harrington, Jonathan. "Evangelicalism, Environmental Activism, and Climate Change in the United States." *Journal of Religion and Society* 11 (2009): 1–24.

Harris, J. Gerald. "A Call to Good Stewardship of Planet Earth." *The Christian Index*, April 10, 2008. www.tciarchive.org/4301.article.

Harvey, Paul. "At Ease in Zion, Uneasy in Babylon: White Evangelicals." In *Religion and Public Life in the South: In the Evangelical Mode*, edited by Charles Reagan Wilson and Mark Silk, 63–78. Walnut Creek, CA: AltaMira Press, 2005.

Hawkins, John. "7 Examples of Discrimination against Christians in America." Townhall, September 17, 2013. https://townhall.com/columnists/johnhawkins /2013/09/17/7-examples-of-discrimination-against-christians-in-america- n1701966.

Hayhoe, Katharine. "Climate, Politics and Religion—My Opinion." *Katharine Hayhoe: Climate Scientist* (blog), June 5, 2015. http://katharinehayhoe.com /wp2016/2015/06/05/climate-politics-and-religion/.

Heartland Institute. "Who We Are." www.heartland.org/about-us/who-we-are. Accessed August 3, 2018.

Heltzel, Peter G. *Jesus and Justice: Evangelicals, Race, and American Politics.* New Haven, CT: Yale University Press, 2009.

Hendricks, Stephenie. *Divine Destruction: Wise Use, Dominion Theology, and the Making of American Environmental Policy.* Hoboken, NJ: Melville House, 2005.

Henneberger, Melinda. "The 2000 Campaign: Writing and Healing; Career in the Balance, Gore Focused His Energy on a Book." *The New York Times,* September 2, 2000. www.nytimes.com/2000/09/03/us/2000-campaign-writing-healing-career-balance-gore-focused-his-energy-book.html.

Hmielowski, Lauren Feldman, Teresa A. Myers, Anthony Leiserowitz, and Edward Maibach. "An Attack on Science? Media Use, Trust in Scientists, and Perceptions of Global Warming." *Public Understanding of Science* 23, no. 7 (2014): 866–883.

Hoekema, Anthony A. *The Bible and the Future.* Grand Rapids, MI: William B. Eerdmans, 1979.

Hooper, John. "Reagan and the 'Sagebrush Rebellion' in Western States." *Chicago Tribune,* October 31, 1980. ProQuest.

Horton, Dave. "Green Distinctions: The Performance of Identity among Environmental Activists." *The Sociological Review* 51, no. 2 (2003): 63–77.

Hulme, Mike. *Why We Disagree about Climate Change: Understanding Controversy, Inaction and Opportunity.* Cambridge: Cambridge University Press, 2009.

Hunter, Joel. *Right Wing, Wrong Bird: Why the Tactics of the Religious Right Won't Fly with Most Conservative Christians.* Longwood, FL: Distributed Church Press, 2006.

Hurd, Dale. "World Leaders Sound Climate Alarm: 'No Time to Waste.'" *CBN News,* November 30, 2015. www1.cbn.com/cbnnews/world/2015 /November/World-Leaders-Tackle-Climate-Change-No-Time-to-Waste.

Inhofe, James. "Catastrophic Global Warming Alarmism Not Based on Objective Science." Speech, Washington DC, July 29, 2003. www.inhofe.senate .gov/epw-archive/press/bsen-inhofe-delivers-major-speech-on-the-science-of-climate-change/b-icatastrophic-global-warming-alarmism-not-based-on-objective-sciencei-ipart-2/i. Accessed April 12, 2017.

———. *The Greatest Hoax: How the Global Warming Conspiracy Threatens Your Future.* Washington, DC: WND Books, 2012.

ISA (Interfaith Stewardship Alliance). "Advisory Board." Internet Archive Wayback Machine, October 24, 2006. https://web.archive.org/web/20061024182135 /http://www.interfaithstewardship.org/pages/advisors.php.

———. "Christian Global Warming Policy News Conference: National Press Club, Tuesday, July 25." Christian Newswire, July 21, 2006. Lexis Nexis.

———. "A Letter to the National Association of Evangelicals on the Issue of Global Warming." www.pbs.org/moyers/moyersonamerica/green/nae_response.pdf. Accessed July 15, 2018.

Jackson, Harry R., Jr., and Tony Perkins. *Personal Faith, Public Policy: The 7 Urgent Issues That We, as People of Faith, Need to Come Together and Solve.* Lake Mary, FL: Frontline, 2008.

Jehovah's Witnesses. "The Earth Has a 'Fever.' Is There a Cure?" Jehovah's Witnesses, September 1, 2008. www.watchtower.org/e/20080901/article_01 .htm (page no longer available).

———. "Will Man Ruin the Earth beyond Repair?" Jehovah's Witnesses, September 1, 2014. www.jw.org/en/publications/magazines/wp20140901/ruined-earth/.

Jenkins, Willis. "After Lynn White: Religious Ethics and Environmental Problems." *Journal of Religious Ethics* 37, no. 2 (June 2009): 283–309.

Jones, Jeffrey M. "In U.S., Concern about Environmental Threats Eases." Gallup, March 25, 2015. www.gallup.com/poll/182105/concern-environmental-threats-eases.aspx.

Jones, Robert E., and Riley Dunlap. "The Social Bases of Environmental Concern: Have They Changed over Time?" *Rural Sociology* 57, no. 1 (March 1992): 28–47.

Jones, Robert P., Daniel Cox, and Juhem Navarro-Rivera. *Believers, Sympathizers and Skeptics: Why Americans Are Conflicted about Climate Change, Environmental Policy and Science.* Washington, DC: Public Religion Research Institute and the American Academy of Religion, 2014.

Kahan, Dan, Donald Braman, Paul Slovic, John Gastil, and Geoffrey Cohen. "The Second National Risk and Culture Study: Making Sense of—and Making Progress in—the American Culture War of Fact." GWU Legal Studies Resesarch Paper no. 370; Yale Law School, Public Law Working Paper no. 154; GWU Law School Public Law Research Paper no. 370; Harvard Law School Program on Risk Regulation Research Paper no. 08–26, October 3, 2007. https://scholarship.law.gwu.edu/cgi/viewcontent.cgi?article=1271&context=faculty_publications.

Kahan, Dan, Ellen Peters, Maggie Wittlin, Paul Slovic, Lisa Larrimore Ouellette, Donald Braman, and Gregory Mandel. "The Polarizing Impact of Science Literacy and Numeracy on Perceived Climate Change Risks." *Nature Climate Change* 2 (2012): 732–735.

Kansas City Star. "Evangelicals Question Assumptions about Global Warming." July 29, 2006. Access World News.

Kaufman, Donald G., and Cecelia M. Franz. *Biosphere 2000: Protecting Our Global Environment.* Dubuque, IA: Kendall Hunt, 2000.

Kearns, Laurel. "Cooking the Truth: Faith, Science, the Market and Global Warming." In *Ecospirit: Religions and Philosophies for the Earth,* edited by Laurel Kearns and Catherine Keller, 97–124. New York: Fordham University Press, 2007.

———. "Green Evangelicals." In *The New Evangelical Social Engagement,* edited by Brian Steensland and Philip Goff, 157–178. Oxford: Oxford University Press, 2014.

———. "Noah's Ark Goes to Washington: A Profile of Evangelical Environmentalism." *Social Compass* 44, no. 3 (1997): 349–366.

———. "Religious Climate Activism in the United States." In *Religion in Environmental and Climate Change: Suffering, Values, Lifestyles,* edited by Dieter Gerten and Sigurd Bergmann, 132–151. London: Continuum, 2012.

Kellstedt, Lyman A., and John C. Green. "Knowing God's Many People: Denominational Preference and Political Behavior." In *Rediscovering the Religious Factor in American Politics,* edited by David C. Leege and Lyman Kellstedt, 53–69. Armonk, NY: M.E. Sharpe, 1993.

Kellstedt, Lyman A., and Corwin E. Smidt. "Measuring Fundamentalism: An Analysis of Different Operational Strategies." In *Religion and the Culture Wars: Dispatches from the Front,* edited by John C. Green, James L. Guth, Corwin E. Smidt, and Lyman A. Kellstedt. Lanham, MD: Rowman & Littlefield, 1996.

Kennedy, D. James, and E. Calvin Beisner. *Overheated: A Reasoned Look at the Global Warming Debate.* Fort Lauderdale, FL: Coral Ridge Ministries Media, 2007.

Kennedy, D. James, and Jerry Newcombe. *How Would Jesus Vote? A Christian Perspective on the Issues.* Colorado Springs, CO: Waterbrook Press, 2008.

Kennedy, Robert F., Jr. *Crimes against Nature: How George W. Bush and His Corporate Pals Are Plundering the Country and Hijacking Our Democracy.* New York: HarperCollins, 2004.

Kilburn, H. Whitt. "Religion and Foundations of American Public Opinion towards Global Climate Change." *Environmental Politics* 23, no. 3 (2014): 473–489.

Kilian, Michael. "Hazy View from Paradise." *Chicago Tribune,* May 23, 1981. ProQuest.

Kilpatrick, James J. "Why Don't They Like James Watt?" *The Washington Post,* August 10, 1981. ProQuest.

Kirkpatrick, David. "The Evangelical Crackup." *The New York Times Magazine,* October 28, 2007. www.nytimes.com/2007/10/28/magazine/28Evangelicals-t.html.

Knight, Graham, and Josh Greenberg. "Talk of the Enemy: Adversarial Framing and Climate Change Discourse." *Social Movement Studies* 10, no. 4 (2011): 323–340.

Krattenmaker, Tom. "A Model of Faith." *USA Today,* June 5, 2006. Access World News.

Krueger, Richard A., and Mary Anne Casey. *Focus Groups: A Practical Guide for Applied Research.* Thousand Oaks, CA: Sage, 2009.

Kruse, Kevin M. *One Nation under God: How Corporate America Invented Christian America.* New York: Basic Books, 2015.

LaHaye, Tim, and David Noebel. *Mind Siege: The Battle for Truth in the New Millennium.* Nashville, TN: Word, 2000.

Lahsen, Myanna. "Climategate: The Role of the Social Sciences." *Climatic Change* 119 (2013): 547–558.

Lampman, Jane. "Southern Baptist Leaders Urge Climate Change Action." *The Christian Science Monitor,* March 12, 2008. www.csmonitor.com

/Environment/Global-Warming/2008/0312/southern-baptist-leaders-urge-climate-change-action.

Land, Richard. "Richard Land Extended Interview." Interview by Bob Abernethy. *Religion & Ethics Newsweekly*, PBS, November 17, 2006. www.pbs .org/wnet/religionandethics/2006/11/17/november-17-2006-richard-land-extended-interview/14063/.

Larsen, David Kenneth. "God's Gardeners: American Protestant Evangelicals Confront Environmentalism, 1967–2000." PhD diss., University of Chicago, 2001.

Lawson, Ronald. "The Persistence of Apocalypticism within a Denominationalizing Sect." In *Millennium, Messiahs, and Mayhem: Contemporary Apocalyptic Movements*, edited by Thomas Robbins and Susan Palmer, 207–208. New York: Routledge, 1997.

Leduc, Timothy B. "Fuelling America's Climatic Apocalypse." *Worldviews* 11, no. 2 (2007): 255–283.

Leggett, Jeremy K. *The Carbon War: Global Warming and the End of the Oil Era*. New York: Routledge, 2001.

Legum, Judd. "Pat Robertson: I'm 'a Convert' on Global Warming, 'It Is Getting Hotter.'" *ThinkProgress*, August 3, 2006. https://thinkprogress.org /pat-robertson-im-a-convert-on-global-warming-it-is-getting-hotter-e59900e f3b0f.

Lehtonen, Mikko. *Cultural Analysis of Texts*. Trans. Aija-Lenna Ahonen and Kris Clarke. London: Sage, 2000.

Leopold, Aldo. *A Sand County Almanac: With Other Essays on Conservation from Round River*. New York: Oxford University Press, 1966 [1949].

Leshy, John D. "Unraveling the Sagebrush Rebellion: Law, Politics and Federal Lands." *U.C. Davis Law Review* 14, no. 2 (1980): 317–355.

Liberty Channel (Lynchburg, VA). "Program Guide." Internet Archive Wayback Machine, January 12, 2005. https://web.archive.org/web/20050112145738 /www.wtlutv.com:80/programguide.html.

Liberty University. "LU Convocation FAQ." www.liberty.edu/faithservice /index.cfm?PID=32826. Accessed July 21, 2017.

Lichterman, Paul. "Religion and the Construction of Civic Identity." *American Sociological Review* 73, no. 1 (2008): 83–104.

Lienesch, Michael. *Redeeming America: Piety and Politics in the New Christian Right*. Chapel Hill: University of North Carolina Press, 1993.

Lindsay, D. Michael. *Faith in the Halls of Power: How Evangelicals Joined the American Elite*. Oxford: Oxford University Press, 2007.

Little, Steve. "A Different View on Global Warming." *CBN News*, September 13, 2008. www.cbn.com/cbnnews/us/2008/September/A-Different-View-on-Global-Warming-/?mobile=false (page no longer available).

Loller, Travis. "Gore Announces Global Warming Effort." Associated Press, March 31, 2008. Access World News.

Longo, Stefano B., and Joseph O. Baker. "Economy 'versus' Environment: The Influence of Economic Ideology and Political Identity on Perceived Threat of Eco-Catastrophe." *The Sociological Quarterly* 55 no. 2 (2014): 341–365.

Lowe, Scott. "Chinese Millennial Movements." In *The Oxford Handbook of Millennialism,* edited by Catherine Wessinger, 307–325. Oxford: Oxford University Press, 2011.

Luo, Michael, and Laurie Goodstein. "Emphasis Shifts for a New Breed of Evangelicals." *The New York Times,* May 21, 2007. www.nytimes.com/2007/05/21/us/21evangelical.html.

Lyons, Eric. "Global Warming, Earth's History, and Jesus' Return." Apologetics Press (website), June 21, 2008. www.apologeticspress.org/APContent.aspx?category=11&article=2521.

Maibach, Edward, Connie Roser-Renouf, and Anthony Leiserowitz. "Global Warming's Six Americas 2009: An Audience Segmentation Analysis." New Haven, CT, and Fairfax, VA: Yale Project on Climate Change and the Georgia Mason University Centre for Climate Change Communication, 2009.

Maier, Harry O. "Green Millennialism: American Evangelicals, Environmentalism, and the Book of Revelation." In *Ecological Hermeneutics: Biblical, Historical, and Theological Perspectives,* edited by David G. Horrell, Cherryl Hunt, Christopher Southgate, and Francesca Stavrakopoulou. London: T&T Clark, 2010.

March, William. "Religious Right at a Low Point." *Tampa Tribune,* March 24, 2007. LexisNexis.

Marsden, George M. *Fundamentalism and American Culture: The Shaping of Twentieth-Century Evangelicalism, 1870–1925.* New York: Oxford University Press, 1980.

———. *Understanding Fundamentalism and Evangelicalism.* Grand Rapids, MI: William B. Eerdmans, 1991.

Martin, William. "Waiting for the End: The Growing Interest in Apocalyptic Prophecy." *The Atlantic Monthly* 249, no. 6 (1982): 31–37.

Mayhew, Elizabeth. "British Advisor Speaks on Climate Change." *Liberty News* (Lynchburg, VA), March 10, 2010. www.liberty.edu/alumni/alumni-news/?MID=16533.

McCammack, Brian. "Hot Damned America: Evangelicalism and the Climate Change Policy Debate." *American Quarterly* 59, no. 3 (2007): 645–668.

McCright, Aaron M. "Political Orientation Moderates Americans' Beliefs and Concern about Climate Change." *Climatic Change* 104, no. 2 (January 2011): 243–253.

McCright, Aaron M., and Riley E. Dunlap. "The Politicization of Climate Change and Polarization in the American Public's Views of Global Warming, 2001–2010." *The Sociological Quarterly* 52 (2011): 155–194.

McCutcheon, Russell T., ed. *The Insider/Outsider Problem in the Study of Religion: A Reader.* London: Continuum, 2005.

McCutcheon, Russell T. *Manufacturing Religion: The Discourse on Sui Generis and the Politics of Nostalgia.* New York: Oxford University Press, 1997.

McDonald, Susan. "Changing Climate, Changing Minds: Applying the Literature on Media Effects, Public Opinion, and the Issue-Attention Cycle to Increase Public Understanding of Climate Change." *International Journal of Sustainability Communication* 4 (2009): 45–63.

McKibben, Bill. "The Gospel of Green: Will Evangelicals Help Save the Earth?" *OnEarth* (Fall 2006): 35–37. www.nrdc.org/onEarth/o6fal/greener1.asp (page no longer available).

———. "A Link between Climate Change and Joplin Tornadoes? Never!" *The Washington Post*, May 23, 2011. www.washingtonpost.com/opinions/a-link-between-climate-change-and-joplin-tornadoes-never/2011/05/23/AFrVC49G_story.html?utm_term=.ca2256c7a22b.

McLaren, Brian. *Everything Must Change: When the World's Biggest Problems and Jesus' Good News Collide.* Nashville, TN: Thomas Nelson, 2007.

Media Matters for America. "Falwell Dismissed Scientific Evidence on Global Warming, Evangelical Efforts to Address Issue." March 14, 2006. http://mediamatters.org/research/2006/03/14/falwell-dismissed-scientific-evidence-on-global/135103.

Merritt, Jonathan. *Green Like God: Unlocking the Divine Plan for Our Planet.* New York: FaithWords, 2010.

———. "Is Mark Driscoll This Generation's Pat Robertson?" *Religion News Service*, May 13, 2013. http://jonathanmerritt.religionnews.com/2013/05/13/is-mark-driscoll-this-generations-pat-robertson/.

Miami Herald. "The Environment in Serious Trouble." December 11, 2004. Access World News.

Miller, David. "Is the Kingdom Yet to Be Established?" Apologetics Press (website), January 1, 2004. www.apologeticspress.org/APContent.aspx?category=11&article=1088&topic=83.

———. *The Silencing of God.* DVD (no date). Montgomery, AL: Apologetics Press.

Mills, Sarah, Barry G. Rabe, and Christopher Borick. "Acceptance of Global Warming Rising for Americans of All Religious Beliefs." Ann Arbor, MI: The Center for Local, State, and Urban Policy at the Gerald R. Ford School of Public Policy, University of Michigan, 2015. http://closup.umich.edu/issues-in-energy-and-environmental-policy/25/acceptance-of-global-warming-among-americans-reaches-highest-level-since-2008/.

Morgan, David L., Richard A. Krueger, and Jean A. King. *Focus Group Kit.* Thousand Oaks, CA: Sage, 1998.

Moser, Susanne C., and Lisa Dilling. "Communicating Climate Change: Closing the Science-Action Gap." In *The Oxford Handbook of Climate Change and Society,* edited by John S. Dryzek, Richard B. Norgaard, and David Schlosberg, 161–174. Oxford: Oxford University Press, 2011.

Movieguide. "NASA Satellites Debunk Global Warming Theory." July 27, 2011. www.movieguide.org/news-articles/nasa-satellites-debunk-global-warming-theory.html.

———. "Our Story." www.movieguide.org/about-movieguide/. Accessed December 4, 2018.

Moyers, Bill. "If We Don't Act Soon, There Is No Tomorrow." *The Oakland Tribune* (Oakland, CA), February 6, 2005. Access World News.

———. "Welcome to Doomsday." *The New York Review of Books* 52 (March 24, 2005). www.nybooks.com/articles/2005/03/24/welcome-to-doomsday/.

———. *Welcome to Doomsday*. New York: The New York Review of Books, 2006.

———. "With God on Their Side." *The Ecologist* 35, no. 4 (2005): 22–24. EBSCO.

NAE (National Association of Evangelicals). "Caring for God's Creation: A Call to Action." Washington, DC: NAE, 2015. www.nae.net/caring-for-gods-creation/.

———. "Ecology." January 1, 1970. Washington, DC: NAE. www.nae.net/ecology/.

———. "Environment and Ecology." January 1, 1971. Washington, DC: NAE. www.nae.net/environment-and-ecology/.

———. "Stewardship." January 1, 1990. Washington, DC: NAE. www.nae.net/stewardship/.

———. "What Is an Evangelical?" Washington, DC: NAE. www.nae.net/what-is-an-evangelical/. Accessed July 18, 2018.

Nagle, John C. "The Evangelical Debate over Climate Change." *University of St. Thomas Law Journal* 5 (2008): 52–86.

Nash, Roderick. *The Rights of Nature: A History of Environmental Ethics*. Madison: University of Wisconsin Press, 1989.

National Religious Broadcasters. "History." http://nrb.org/about/history/. Accessed July 16, 2018.

Nisbet, Matthew. "Communicating Climate Change: Why Frames Matter for Public Engagement." *Environment* 51 no. 2 (2009): 12–23.

Norgaard, Kari Marie. *Living in Denial: Climate Change, Emotions, and Everyday Life*. Cambridge, MA: MIT Press, 2011.

Northcott, Michael S. "The Dominion Lie: How Millennial Theology Erodes Creation Care." In *Diversity and Dominion: Dialogues in Ecology, Ethics and Theology*, edited by K.S.V. Houtan and M.S. Northcott, 89–108. Eugene, OR: Wipf and Stock, 2010.

Oard, Michael J. "Is Man the Cause of Global Warming?" In *The New Answers Book 3: Over 35 Questions on Creation/Evolution and the Bible*, vol. 3, edited by Ken Ham, 69–80. Green Forest, AR: Master Books, 2009.

Omang, Joanne. "Watt, a Man with a Mission, Launches New Era at Interior." *The Washington Post*, March 9, 1981. LexisNexis.

Oreskes, Naomi. "Religion and Climate Change Skepticism." Paper presented at the annual meeting of the American Academy of Religion, San Diego, CA, November 24, 2014.

———. "The Scientific Consensus on Climate Change." *Science* 306, no. 5702 (December 2004): 1686.

Oreskes, Naomi, and Erik M. Conway. *Merchants of Doubt: How a Handful of Scientists Obscured the Truth on Issues from Tobacco Smoke to Global Warming*. New York: Bloomsbury Perss, 2010.

Orr, David W. "Armageddon versus Extinction." *Conservation Biology* 19, no. 2 (November 2005): 290–292.

———. *Ecological Literacy: Education and the Transition to a Postmodern World*. Albany: State University of New York Press, 1992.

Ostling, Richard. "Power, Glory—and Politics." *Time,* June 24, 2001. http://content.time.com/time/magazine/article/0,9171,143137,00.html.

Page, Frank. "Frank Page Statement on Climate Change." *Baptist Press,* March 11, 2008. www.bpnews.net/bpnews.asp?id=27601.

———. "Spirited and Christ-Like Climate Debate." *The Christian Index,* April 10, 2008. www.tciarchive.org/4299.article.

Palmer, Clare. "An Overview of Environmental Ethics." In *Environmental Ethics: An Anthology,* edited by Andrew Light and Holmes Rolston III, 15–37. Malden, MA: Blackwell, 2003.

Pasztor, Andy. "James Watt Tackles Interior Agency Job with Religious Zeal." *The Wall Street Journal,* May 5, 1981. ProQuest.

Peifer, Jared L., Elaine Howard Ecklund, and Cara Fullerton. "How Evangelicals from Two Churches in the American Southwest Frame Their Relationship with the Environment." *Review of Religious Research* 56, no. 3 (September 2014): 373–397.

Perkins, Tony. "Climate Change." Family Research Council (radio commentary), November 18, 2015. www.frc.org/washingtonwatchdailyradiocommentary/climate-change.

———. "If NAE's Rich Cizik Doesn't Speak for Them, Who Does He Speak For?" *FRC Blog,* December 10, 2008. www.frcblog.com/2008/12/if-naes-rich-cizik-doesnt-speak-for-them-who-does-he-speak-for/.

———. "Introduction." In *25 Pro-Family Policy Goals for the Nation,* edited by Peter Sprigg. Washington, DC: Family Research Council, 2008. https://downloads.frc.org/EF/EF11D02.pdf.

———. "NAE's Dangerous Emissions on Global Warming." FRC Blog, March 6, 2007. www.frcblog.com/2007/03/naes-dangerous-emissions-on-global-warming/.

Peterson, Cass. "Executive Notes." *The Washington Post,* February 18, 1981, A3. LexisNexis.

Pew Research Center. "Growing Numbers of Americans Say Obama Is a Muslim." August 18, 2010. Washington, DC: Pew Research Center. www.pewforum.org/2010/08/18/growing-number-of-americans-say-obama-is-a-muslim/.

———. "Jesus Christ's Return to Earth." July 14, 2010. Washington, DC: Pew Research Center. http://pewrsr.ch/T4uCxO.

———. "The Political Preference of U.S. Religious Groups." February 23, 2016. Washington, DC: Pew Research Center. www.pewresearch.org/fact-tank/2016/02/23/u-s-religious-groups-and-their-political-leanings/ft_16–02–22_religionpoliticalaffiliation_640px-2/.

———. "Protecting the Environment Ranks in the Middle of Public's Priorities for 2013." April 22, 2013. Washington, DC: Pew Research Center. www.pewresearch.org/daily-number/protecting-the-environment-ranks-in-the-middle-of-publics-priorities-for-2013/.

———. "Science in America: Religious Belief and Public Attitudes." December 18, 2007. www.pewforum.org/2007/12/18/science-in-america-religious-belief-and-public-attitudes/.

———. "The Tea Party and Religion." February 23, 2011. Washington, DC: Pew Research Center. www.pewforum.org/2011/02/23/tea-party-and-religion/.

———. "U.S. Public Becoming Less Religious." November 23, 2015. Washington, DC: Pew Research Center. www.pewforum.org/2015/11/03/chapter-4-social-and-political-attitudes/.

———. "U.S. Religious Landscape Survey." 2008. Washington, DC: Pew Research Center, Religion and Public Life Project. www.pewforum.org/religious-landscape-study/.

———. "Views about Abortion among Evangelical Protestants by Religious Group." 2014. Washington, DC: Pew Research Center. www.pewforum.org/religious-landscape-study/compare/views-about-abortion/by/religious-family/among/religious-tradition/evangelical-protestant/.

Pew Research Center/Council on Foreign Relations. "America's Place in the World IV, Oct. 2005." Poll conducted by Princeton Survey Research Associates, October 12–24, 2005. Roper Center for Public Opinion Research, Cornell University, Ithaca, NY. https://ropercenter.cornell.edu/. Accessed December 12, 2018.

Phillips, Kevin. *American Theocracy: The Peril and Politics of Radical Religion, Oil, and Borrowed Money in the 21st Century.* New York: Penguin Group, 2006.

Pierson, Russ, and John Roe. "Mark Driscoll: Gas-Guzzlers a Mark of Masculinity." *Sojourners,* May 9, 2013. https://sojo.net/articles/mark-driscoll-gas-guzzlers-mark-masculinity.

PIPA (Program on International Policy Attitudes). "Americans on Addressing World Poverty, the Crisis in Darfur and US Trade, Jun. 2005." Poll conducted by Knowledge Networks, June 22–26, 2005. Roper Center for Public Opinion Research, Cornell University, Ithaca, NY. https://ropercenter.cornell.edu/. Accessed December 12, 2018.

Poloma, Margaret M., and John C. Green. *The Assemblies of God: Godly Love and the Revitalization of American Protestantism.* New York: New York University Press, 2010.

Pope, Carl. 2007. "Riding the Big One." *Sierra* 92 no. 4 (July/August 2007): 6. Energy and Power Source (EBSCO).

Prentice, David. "Change Watch Backgrounder: Jane Lubchenco." Family Research Council, February 12, 2009. www.frcblog.com/2009/02/change-watch-backgrounder-jane-lubchenco/.

Prochnau, Bill. "The Watt Controversy: Interior Secretary Watt Creates Political Stresses for Administration." *The Washington Post,* June 30, 1981. LexisNexis.

PRRI (Public Religion Research Institute) and Brookings Institute. "How Immigration and Concerns about Cultural Changes Are Shaping the 2016 Election." Washington, DC: Public Religion Research Institute, 2016. www.prri.org/wp-content/uploads/2016/06/PRRI-Brookings-2016-Immigration-survey-report.pdf.

PRRI and RNS (Religion News Service). "Few Americans See Earthquakes, Floods and Other Natural Disasters as a Sign from God." March 24, 2011.

http://publicreligion.org/research/2011/03/few-americans-see-earthquakes-floods-and-other-natural-disasters-a-sign-from-god-2/.

Pyle, Nate. "Why Mark Driscoll's Theology of SUV's Matters." *Nate Pyle* (blog), May 3, 2013. http://natepyle.com/why-mark-driscolls-theology-of-suvs-matters/.

Rabe, Barry G., and Christopher Borick. *National Surveys on Energy and the Environment* (United States) (Spring 2010, Fall 2010, Spring 2013, Fall 2013, Spring 2014, Fall 2014, Spring 2015, Fall 2015 datasets). Ann Arbor, MI: Inter-university Consortium of Political and Social Research, 2017 (Muhlenberg Institute of Public Opinion at Muhlenberg College and the Center for Local, State and Urban Policy at the University of Michigan's Gerald R. Ford School of Public Policy).

Rabe, John. "Christian Views on Global Warming." Coral Ridge Ministries, April 7, 2008. www.truthcasting.com/Coral-Ridge-Ministries-Ft.-Lauderdale-FL-Christian-Views-On-Global-Warming-841.sermon.

Radway, Janice. *Reading the Romance: Women, Patriarchy, and Popular Literature.* Chapel Hill: University of North Carolina Press, 1984.

Ratio Christi. "Calvin Beisner Personal Profile." http://ratiochristi.org/people/calvin-beisner. Accessed June 15, 2016 (page no longer available).

Ray, Janisse. *Ecology of a Cracker Childhood.* Minneapolis, MN: Milkweed Editions, 1999.

———. *Wild Card Quilt: The Ecology of Home.* Minneapolis, MN: Milkweed Editions, 2003.

Reed, Nathaniel Pryor. "In the Matter of Mr. Watt . . ." *Sierra: The Sierra Club Bulletin* 66, no. 4 (1981): 6–15.

Religion and Ethics Newsweekly and U.S. News and World Report. "America's Evangelicals." Poll conducted by Greenberg Quinlan Rosner Research, March 16–April 4, 2004. Roper Center for Public Opinion Research, Cornell University, Ithaca, NY. https://ropercenter.cornell.edu/. Accessed December 12, 2018.

Religion News Service. "Posthumous Preaching." *Christianity Today* 53, no. 6 (2009): 16.

Research & Polling Inc. "Focus Group Report: Perceptions of Global Warming." Albuquerque, NM: Research & Polling, 2006.

Riley, Matthew T. "The Democratic Roots of Our Ecologic Crisis: Lynn White, Biodemocracy and the Earth Charter." *Zygon* 49, no. 4 (December 2014): 938–948.

Roach, David. "End Times: Scholars Differ on What the Bible Says about Subject." *Baptist Press* (Nashville), December 30, 2009. www.bpnews.net/BPnews.asp?ID=31963.

Roberts, Jon H. *Darwinism and the Divine in America: Protestant Intellectuals and Organic Evolution, 1859–1900.* Madison: University of Wisconsin Press, 1988.

Roberts, Michael. "Evangelicals and Climate Change." In *Religion in Environmental and Climate Change,* edited by Dieter Gerten and Sigurd Bergmann, 107–131. London: Continuum, 2012.

Robertson, Pat. *The New World Order.* Dallas: Word, 1991.

Ronan, Marisa. "American Evangelicalism, Apocalypticism, and the Anthropocene." In *Religion in the Anthropocene,* edited by Celia Deane-Drummond, Sigurd Bergmann, and Markus Vogt, 218–232. Eugene, OR: Wipf and Stock, 2017.

Roser-Renouf, Connie, Edward Maibach, Anthony Leiserowitz, and Seth Rosenthal. *Global Warming, God, and the "End Times."* Yale University and George Mason University. New Haven, CT: Yale Program on Climate Change Communication, 2016. http://climatecommunication.yale.edu /publications/global-warming-god-end-times/.

Rossing, Barbara. *The Rapture Exposed: The Message of Hope on the Book of Revelation.* New York: Basic Books, 2004.

Rutledge, Josh. "Evangelicals Spar over Climate—Alliance Slams Global-Warming Call to Action." *The Washington Times,* July 31, 2006. Access World News.

Salem Web Network. "Advertising." Internet Archive Wayback Machine, September 29, 2007. http://web.archive.org/web/20070929160540/www .salemwebnetwork.com:80/advertising.asp.

Saunders, Bill. "An Inconvenient Ambiguity: Evangelicals Should Not Be Fooled by Global-Warming Hysterics." *Human Events,* November 6, 2006, 9.

SBC (Southern Baptist Convention). "The Baptist Faith and Message." www.sbc .net/bfm/bfm2000.asp. Accessed January 12, 2012 (page no longer available).

———. "On Environmentalism and Evangelicals." Greensboro, NC: Southern Baptist Convention, 2006. www.sbc.net/resolutions/1159/on-environmentalism-and-evangelicals.

———. "On Global Warming." San Antonio: Southern Baptist Convention, 2007. www.sbc.net/resolutions/1171/on-global-warming.

———. "Resolution on Environmental Stewardship." New Orleans: Southern Baptist Convention, 1990. www.sbc.net/resolutions/456/resolution-on-environmental-stewardship.

SBECI (Southern Baptist Environment and Climate Initiative). "A Southern Baptist Declaration on the Environment and Climate Change." March 10, 2008. www.baptistcreationcare.org/node/1 (site discontinued).

Schaeffer, Francis A. *Pollution and the Death of Man: The Christian View of Ecology.* Wheaton, IL: Tyndale, 1970.

Schell, Jonathan. "Talk of the Town: Notes and Comment." *The New Yorker,* May 18, 1981.

Scherer, Glenn. "Christian-Right Views Are Swaying Politicians and Threatening the Environment." *Grist* (Seattle, WA), October 27, 2004. www.grist .org/news/maindish/2004/10/27/scherer-christian/.

———. "George Bush's War on Nature." *Salon,* January 6, 2003. www.salon .com/2003/01/06/nature/.

———. "Religious Wrong: A Higher Power Informs the Republican Assault on the Environment." *E-The Environmental Magazine,* [May/June, 2003?]. https:// emagazine.com/religious-wrong/ (originally acccessed at www.emagazine.com /magazine-archive/religious-wrong, page no longer available).

———. "Why Ecocide Is 'Good News' for the GOP." AlterNet, May 4, 2003. http://web.alternet.org/story/15814/why_ecocide_is_%27good_news%27_ for_ the_gop (page no longer available).

Schlozman, Daniel. *When Movements Anchor Parties: Electoral Alignments in American History*. Princeton, NJ: Princeton University Press, 2015.

Schreibman, Jack. "Sierra Club Criticizes Interior Secretary." Associated Press, April 3, 1981, AM Cycle. Nexis Uni.

Seventh-day Adventist Church. "28 Fundamental Beliefs." Silver Spring, MD: General Conference of Seventh-day Adventists. www.adventist.org/beliefs/. Accessed November 5, 2013 (page no longer available).

———. "The Dangers of Climate Change." Costa Rica: General Conference of Seventh-day Adventists Administrative Committee, December 19, 1995. www.adventist.org/en/information/official-statements/statements/article/go/-/the-dangers-of-climate-change/.

———. "Environment." Utrecht, Netherlands: General Conference of Seventh-day Adventists Administrative Committee, June 29, 1995. www.adventist.org/en/information/official-statements/statements/article/go/-/environment-1/.

———. "Glimpses of Our God." *Adult Sabbath School Bible Study Guide*, January–March 2012. Silver Spring, MD: General Conference Administrative Committee.

———. "Stewardship of the Environment." Costa Rica: General Conference of Seventh-day Adventists Administrative Committee, October 1, 1996. www.adventist.org/en/information/official-statements/statements/article/go/o/stewardship-of-the-environment/.

Severson, Kim, and Kirk Johnson. "Drought Spreads Pain from Florida to Arizona." *The New York Times*, July 11, 2011. www.nytimes.com/2011/07/12/us/12drought.html?_r=0.

Shabecoff, Philip. *A Fierce Green Fire: The American Environmental Movement*. Washington, DC: Island Press, 2003.

Shaiko, Ronald G. "Religion, Politics, and Environmental Concern: A Powerful Mix of Passions." *Social Science Quarterly* 68, no. 2 (June 1987): 244–262.

Shao, Wanyun. "Weather, Climate, Politics, or God? Determinants of American Public Opinions toward Global Warming." *Environmental Politics* 26, no. 1 (2017): 71–96.

Sherkat, D.E., and C.G. Ellison. "Structuring the Religion-Environment Connection: Identifying Religious Influences on Environmental Concern and Activism." *Journal for the Scientific Study of Religion* 46, no. 1 (March 2007): 71–85.

Shibley, Mark A. *Resurgent Evangelicalism in the United States: Mapping Cultural Change since 1970*. Columbia, SC: University of South Carolina Press, 1996.

———. "Sacred Nature: Earth-Based Spirituality as Popular Religion in the Pacific Northwest." *Journal for the Study of Religion, Nature and Culture* 5, no. 2 (2011): 164–185.

Sider, Ron. "ESA's History: A Reflection." Evangelicals for Social Action (website). www.evangelicalsforsocialaction.org/about/history/. Accessed February 17, 2014.

Simmons, J. Aaron. "Evangelical Environmentalism: Oxymoron or Opportunity?" *Worldviews* 13, no. 1 (2009): 40–71.

Skeel, David A. "Evangelicals, Climate Change, and Consumption." *Environmental Law Reporter* 38 (2008): 10868–10872.

Skocpol, Theda. "Naming the Problem: What It Will Take to Counter Extremism and Engage Americans in the Fight against Global Warming." Report prepared for the Symposium on the Politics of America's Fight against Global Warming, Harvard University, Boston, MA, 2013.

Sky Angel. "Sky Angel's Angel One National Network." Internet Archive Wayback Machine, May 17, 2007. https://web.archive.org/web/20070517072337/www.skyangel.com:80/About/Index.asp?Reference=AirtimeSales&~=.

Slimak, Michael W., and Thomas Dietz. "Personal Values, Beliefs, and Ecological Risk Perception." *Risk Analysis* 26, no. 6 (2006): 1689–1705.

Smith, Amy Erica, and Robin Globus Veldman. "Conservative Protestant Environmentalists? Evidence from Brazil." Manuscript in preparation.

Smith, Christian, Michael Emerson, Sally Gallagher, Paul Kennedy, and David Sikkink. *American Evangelicalism: Embattled and Thriving.* Chicago: University of Chicago Press, 1998.

Smith, E. Keith, Lynn M. Hempel, and Kelsey MacIlroy. "What's 'Evangelical' Got to Do with It? Disentangling the Impact of Evangelical Protestantism on Environmental Outcomes." *Environmental Politics* 27, no. 2 (2018): 292–319.

Smith, N., and A. Leiserowitz. "American Evangelicals and Global Warming." *Global Environmental Change* 23, no. 5 (2013): 1009–1017.

Spencer, Roy W., Paul K. Driessen, and E. Calvin Beisner. "An Examination of the Scientific, Ethical and Theological Implications of Climate Change Policy." Interfaith Stewardship Alliance, 2005. www.cornwallalliance.org/docs/an-examination-of-the-scientific-ethical-and-theological-implications-of-climate-change-policy.pdf.

Stark, Rodney, and William S. Bainbridge. *The Future of Religion: Secularization, Revival, and Cult Formation.* Berkeley: University of California Press, 1985.

Stark, Rodney, and Charles Y. Glock. "Prejudice and the Churches." In *Prejudice U.S.A.*, edited by Charles Y. Glock and Ellen Siegelman, 70–95. New York: Praeger, 1969.

St. Clair, Jeffrey. "Santorum: That's Latin for Asshole." *Counterpunch,* April 30, 2003. www.counterpunch.org/2003/04/30/santorum/.

Steed, Jason Paul. "Joke-Making Jews/Jokes Making Jews: Essays on Humor and Identity in American Jewish Fiction." PhD diss., University of Nevada, Las Vegas, 2004.

Steinfels, Peter. "Beliefs; in a Wide-Ranging Talk, Al Gore Reveals the Evangelical and Intellectual Roots of His Faith." *The New York Times,* May 29, 1999. www.nytimes.com/1999/05/29/us/beliefs-wide-ranging-talk-al-gore-reveals-evangelical-intellectual-roots-his.html.

Stephens, Randall, and Karl Giberson. *The Anointed: Evangelical Truth in a Secular Age.* Cambridge, MA: Harvard University Press, 2011.

Stewart, Katherine. "The New Anti-science Assault on US Schools." *The Guardian,* February 12, 2012. www.theguardian.com/commentisfree/cifamerica/2012/feb/12/new-anti-science-assault-us-schools.

Stiglitz, D. Joseph. "Moving beyond Market Fundamentalism to a More Balanced Economy." *Annals of Public and Cooperative Economics* 80, no. 3 (September 2009): 345–360.

Stoll, Mark. *Inherit the Holy Mountain: Religion and the Rise of American Environmentalism.* Oxford: Oxford University Press, 2015.

———. "Sinners in the Hands of an Ecologic Crisis: Lynn White's Environmental Jeremiad." In *Religion and the Ecological Crisis: The "Lynn White Thesis" at Fifty,* edited by Todd LeVasseur and Anna Peterson, 47–60. New York: Routledge, 2017.

Stonestreet, John. "The Paris Accord and Climate Change: Salvation by Technocracy?" *BreakPoint,* December 15, 2015. www.breakpoint.org/2015/12/paris-accord-climate-change/.

Strandberg, Todd. "Bible Prophecy and Environmentalism." RaptureReady. Internet Archive Wayback Machine, January 12, 2016. https://web.archive.org/web/20161121120432/www.raptureready.com:80/rr-environmental.html.

Strauss, Anselm L., and Juliet M. Corbin. *Basics of Qualitative Research: Techniques and Procedures for Developing Grounded Theory,* 2nd ed. Thousand Oaks, CA: Sage, 1998.

Strupp, Joe. "Bill Moyers Apologizes to James Watt for Apocryphal Quote." *Editor & Publisher,* February 9, 2005. www.editorandpublisher.com/news/bill-moyers-apologizes-to-james-watt-for-apocryphal-quote/.

Subramanian, Meera. "Generation Climate: Can Young Evangelicals Change the Climate Debate?" *Inside Climate News,* November 21, 2018. https://insideclimatenews.org/news/21112018/evangelicals-climate-change-action-creation-care-wheaton-college-millennials-yeca.

Sullivan, Ned. "Messing with God's Creation: The New Religious Activism to Save the Environment." *E-The Environmental Magazine* 17, no. 3 (May/June 2006): 15–18. MasterFILE Premier.

Sumser, John. "Conservative Talk Radio, Religious Style: When You Need Some Moral Outrage." In *The Electronic Church in the Digital Age: Cultural Impacts of Evangelical Mass Media,* vol. 2, edited by Mark Ward Sr., 105–130. Santa Barbara, CA: ABC-CLIO, 2016.

Sutton, Matthew A. *American Apocalypse: A History of Modern Evangelicalism.* Cambridge, MA: Harvard University Press, 2014.

Swartz, David. *Moral Minority: The Evangelical Left in an Age of Conservatism.* Philadelphia: University of Pennsylvania Press, 2012.

Swidler, Anne. "Culture in Action: Symbols and Strategies." *American Sociological Review* 51, no. 2 (April 1986): 273–286.

Tashman, Brian. "James Inhofe Says the Bible Refutes Climate Change." *Right Wing Watch,* March 8, 2012. www.rightwingwatch.org/post/james-inhofe-says-the-bible-refutes-climate-change/.

Taylor, Bron. "Critical Perspectives on 'Religions of the World and Ecology.'" In *The Encyclopedia of Religion and Nature,* edited by Bron Taylor, 1375–1376. London: Thoemmes Continuum, 2005.

————. *Dark Green Religion: Nature Spirituality and the Planetary Future.* Berkeley: University of California Press, 2010.

————. "Toward a Robust Scientific Investigation of the 'Religion' Variable in the Quest for Sustainability." *Journal for the Study of Religion, Nature and Culture* 5, no. 3 (2011): 253–262.

Taylor, Bron, Gretel Van Wieren, and Bernard Zaleha. "Lynn White, Jr. and the Greening-of-Religion Hypothesis." *Conservation Biology* 30, no. 5 (2016): 1000–1009.

Terris, Ben. "Jim Inhofe Is a Small-Plane-Flying, Global-Warming-Denying Senator. And Now He's Got a Gavel." *The Washington Post,* January 8, 2015. www.washingtonpost.com/lifestyle/style/the-senates-top-climate-change-denier-is-flying-high-with-a-committee-chairmanship/2015/01/07/bf625feo-9365-11e4-ba53-a477d66580ed_story.html?utm_term=.c427d3c70460.

Thomashow, Mitchell. *Bringing the Biosphere Home: Learning to Perceive Global Environmental Change.* Cambridge, MA: MIT Press, 2001.

Thompson, Damian. *Waiting for Antichrist: Charisma and Apocalypse in a Pentecostal Church.* Oxford: Oxford University Press, 2005.

Throckmorton, Warren, and Michael Coulter. *Getting Jefferson Right: Fact Checking Claims about Our Third President.* Grove City, PA: Salem Grove, 2012.

Tierney, John. "And on the Eighth Day, God Went Green." *The New York Times,* February 11, 2006. Access World News.

Time. "The 25 Most Influential Evangelicals in America—David Barton." February 7, 2005. http://content.time.com/time/specials/packages/article/0,28804,1993235_1993243_1993261,00.html.

————. "The 25 Most Influential Evangelicals in America—Rick Warren." February 7, 2005. http://content.time.com/time/specials/packages/article/0,28804,1993235_1993243_1993257,00.html.

Tooley, Mark. "Nothing New under the Sun." *The American Spectator,* July 31, 2006. LexisNexis.

Trollinger, Susan L., and William Vance Trollinger. *Righting America at the Creation Museum.* Baltimore: Johns Hopkins University Press, 2016.

Truesdale, Al. "Last Things First: The Impact of Eschatology on Ecology." *Perspectives on Science and Christian Faith* 46, no. 2 (1994): 116–120.

TruthsThatTransform.org. Home page. Internet Archive Wayback Machine, April 29, 2007. http://web.archive.org/web/20070429131758/http://truthsthattransform.org/.

Tucker, Mary Evelyn. "The Emerging Alliance of Religion and Ecology." *Worldviews* 1, no. 1 (January 1997): 3–34.

United States Congress. Joint Committee on Printing. "2015–2016 Official Congressional Directory, 114th Congress." United States Congress Joint Committee on Printing. Washington, DC: U.S. Government Publishing Office, 2016.

University Advancement Staff. "Ben Stein to Speak at May Commencement." *Liberty News* (Lynchburg, VA), April 6, 2009. www.liberty.edu/news/index.cfm?PID=18495&MID=6873.

Unruh, Bob. "Attacks on Christians in U.S. Double in Three Years." WND, February 22, 2016. www.wnd.com/2016/02/attacks-on-christians-in-america-double-in-3-years/.

Van Biema, David. "The Greening of the Baptists." *Time,* March 10, 2008. www.time.com/time/nation/article/0,8599,1721132,00.html.

Vanderheiden, Steve. 2008. *Atmospheric Justice: A Political Theory of Climate Change.* Oxford: Oxford University Press, 2008.

VCY America Ministry. "About the VCY America Ministry." Milwaukee, MN: VCY America. www.vcyamerica.org/about.htm. Accessed April 12, 2017.

Veldman, Robin Globus. "What Is the Meaning of Greening? Cultural Analysis of a Southern Baptist Environmental Text." *Journal of Contemporary Religion* 31, no. 2 (May 2016): 199–222.

Veldman, Robin Globus, Andrew Szasz, and Randolph Haluza-DeLay. *How the World's Religions Are Responding to Climate Change: Social Scientific Investigations.* Abingdon, UK: Routledge, 2014.

———. "Introduction: Climate Change and Religion—A Review of Existing Research." *Journal for the Study of Religion, Nature and Culture* 6, no. 3 (2012): 255–275.

Victory FM (Lynchburg, VA). "Programs." Internet Archive Wayback Machine, January 1, 2007. https://web.archive.org/web/20070101090232/www.wrvlfm .com:80/index.cfm?PID=8636.

Viguerie, Richard. *The New Right: We're Ready to Lead.* Falls Church, VA: Viguerie, 1981.

Vu, Michelle. "Christians Launch Campaign against Global Warming Hype." *The Christian Post,* May 16, 2008. www.christianpost.com/news/christians-launch-campaign-against-global-warming-hype-32396/.

Wacker, Grant. *Heaven Below: Early Pentecostals and American Culture.* Cambridge, MA: Harvard University Press, 2001.

Wald, Kenneth D., and Allison Calhoun-Brown. *Religion and Politics in the United States,* 6th ed. Lanham, MD: Rowman & Littlefield, 2011.

Wald, Kenneth D., Dennis E. Owen, and Samuel S. Hill. "Churches as Political Communities." *The American Political Science Review* 82, no. 2 (1988): 531–548.

WallBuilders Live. "Stations." www.wallbuilderslive.com/stations.asp. Accessed June 3, 2016 (page no longer available).

Wallis, Jim, and Wes Michaelson. "The Plan to Save America." *Sojourners* (April 1976). ATLA Religion Database.

Ward, Mark, Sr. "Appendix B: Major Networks and Personalities." In *The Electronic Church in the Digital Age: Cultural Impacts of Evangelical Mass Media,* vol. 1, edited by Mark Ward Sr., 255–284. Santa Barbara, CA: ABC-CLIO, 2016.

———. "Give the Winds a Mighty Voice: Evangelical Culture as Radio Ecology." *Journal of Radio and Audio Media* 21, no. 1 (2014): 115–133.

———. "Introduction." In *The Electronic Church in the Digital Age: Cultural Impacts of Evangelical Mass Media,* vol. 1, edited by Mark Ward Sr., xvii-xxvii. Santa Barbara, CA: ABC-CLIO, 2016.

Wardekker, J. Arjan, Arthur C. Petersen, and Jeroen P. van der Sluijs. "Ethics and Public Perception of Climate Change: Exploring the Christian Voices in the US Public Debate." *Global Environmental Change* 19, no. 4 (October 2009): 512–521.

Waskey, Andrew. "Religion." In *Encyclopedia of Global Warming and Climate Change,* edited by S. George Philander, 855–857. Thousand Oaks, CA: Sage, 2008.

Watt, James. *The Courage of a Conservative.* New York: Simon and Schuster, 1985.

———. "The Religious Left's Lies." *The Washington Post,* May 21, 2005.

Watt, Leilani. *Caught in the Conflict: My Life with James Watt.* Jackson Hole, WY: Words to Encourage, 1991.

Weber, Eugen. *Apocalypses: Prophecies, Cults, and Millennial Beliefs through the Ages.* Cambridge, MA: Harvard University Press, 2000.

Weber, Timothy P. "Millennialism." In *The Oxford Handbook of Eschatology,* edited by Jerry L. Walls, 365–383. Oxford: Oxford University Press, 2008.

Westbury, Joe. "Jonathan Merritt: In His Own Words." *The Christian Index,* April 10, 2008. www.tciarchive.org/4331.article.

———. "Seminary President Lauds Student's Courage." *The Christian Index,* April 10, 2008. www.tciarchive.org/4327.article.

———. "Younger Conservative Leaders Need a Voice in SBC." *The Christian Index,* April 10, 2008. www.tciarchive.org/4329.article.

Weyrich, Paul. "A Falst [sic] Frenzy on Global Warming." *Townhall,* July 9, 2008. https://townhall.com/columnists/paulweyrich/2008/07/09/a-falst-frenzy-on-global-warming-n811980.

White, Alan. "Should We Be Concerned about Climate Change?" In *The New Answers Book 4: Over 30 Questions on Creation/Evolution and the Bible,* edited by Ken Ham, 187–198. Green Forest, AR: Master Books, 2013.

White, Lynn, Jr. "The Historical Roots of Our Ecologic Crisis." *Science* 155, no. 3767 (1967): 1203–1207.

Whitehead, Andrew. "Christian Nationalism." In *American Values, Mental Health, and Using Technology in the Age of Trump: Findings from the Baylor Religion Survey, Wave 5,* 18–23. Waco, TX: Baylor Religion Surveys and the Institute for the Studies of Religion, 2017. www.baylor.edu/Baylor ReligionSurvey.

Wilkinson, Katharine K. *Between God and Green: How Evangelicals Are Cultivating a Middle Ground on Climate Change.* Oxford: Oxford University Press, 2012.

———. "Climate's Salvation: Why and How American Evangelicals Are Engaging with Climate Change." *Environment* 52, no. 2 (2010): 47–57.

Williams, Daniel K. *God's Own Party: The Making of the Christian Right.* Oxford: Oxford University Press, 2010.

Williams, Rhys H. "Politicized Evangelicalism and Secular Elites: Creating a Moral Other." In *Evangelicals and Democracy in America: Religion and Politics,* vol. 2, edited by Steven Brint and Jean Reith Schroedel, 143–178. New York: Russell Sage Foundation, 2009.

Wilson, Bruce. "Pat Robertson's Sweaty Global Warming Epiphany Challenges American Environmental Movement." Talk to Action, August 5, 2006. www.talk2action.org/story/2006/8/5/123817/6229.

Wilson, Edward O. *Consilience: The Unity of Knowledge.* New York: Alfred A. Knopf, 1998.

———. *The Creation: An Appeal to Save Life on Earth*. New York: Norton, 2006.

WND. "About WND." Accessed April 5, 2017. http://go.wnd.com/aboutwnd.

WNG Media. "2016 Media Planning Guide." https://world.wng.org/sites /default/files/WorldMediaGuideDigital.pdf. Accessed December 4, 2018.

Wojcik, Daniel. *The End of the World as We Know It: Faith, Fatalism, and Apocalypse in America*. New York: New York University Press, 1997.

Wolf, Ron. "God, James Watt, and the Public's Land." *Audubon*, May 1981, 58–65.

Wood, James. "Senator Santorum's Planet." *The New Yorker*, February 23, 2012. www.newyorker.com/news/daily-comment/senator-santorums-planet.

Woodberry, Robert D., and Christian S. Smith. "Fundamentalism et al.: Conservative Protestantism in America." *Annual Review of Sociology* 24 (1998): 25–56.

Woodruff, Judy. *Inside Politics* (television broadcast). CNN, November 20, 2002. http://transcripts.cnn.com/TRANSCRIPTS/0211/20/ip.00.html.

Worthen, Molly. "Who Would Jesus Smack Down?" *The New York Times*, January 6, 2009. www.nytimes.com/2009/01/11/magazine/11punk-t.html.

Wrench, Jason. "Setting the Evangelical Agenda: The Role of 'Christian Radio.'" In *The Electronic Church in the Digital Age*, vol. 1, edited by Mark Ward Sr., 173–192. Santa Barbara, CA: ABC-CLIO, 2016.

Wright, Richard. "Tearing Down the Green: Environmental Backlash in the Evangelical Subculture." *Perspectives on Science and Christian Faith* 47 (1995): 80–91.

Zaleha, Bernard D. "A Tale of Two Christianities: The Religiopolitical Clash over Climate Change within America's Dominant Religion." PhD diss., University of California, Santa Cruz, 2018.

Zaleha, Bernard D., and Andrew Szasz. "Keep Christianity Brown! Climate Denial on the Christian Right in the United States." In *How the World's Religions Are Responding to Climate Change: Social Scientific Investigations*, edited by Robin Veldman, Andrew Szasz, and Randolph Haluza-DeLay, 209–228. Abingdon, UK: Routledge, 2014.

Zerubavel, Eviatar. *Social Mindscapes: An Invitation to Cognitive Sociology*. Cambridge, MA: Harvard University Press, 1997.

Zoll, Rachel. "Southern Baptist Leaders Take Unusual Step of Urging Fight against Climate Change." Associated Press, March 10, 2008. Access World News.

Index

Page references in italics indicate an illustration, and those followed by a *t* indicate a table.

by, 5; Ethics and Religious Liberty Commission (ERLC), 135–36, 145, 199, 269n81; on gay marriage, 145; in Georgia, 138; on global warming, 132 (*see also* SBECI; "A Southern Baptist Declaration on the Environment and Climate Change"); methodology for studying climate skeptics in, 226–27; "On Environmentalism and Evangelicals," 150, 257n50; progressives (moderates) vs. fundamentalists within, 255n35; religious and political conservatism of, 131–32; Republican vs. Democratic members of, 4; "Resolution on Environmental Stewardship," 150; and responsibility to be "salt and light," 145; size of, 132

"A Southern Baptist Declaration on the Environment and Climate Change," 5, 132–56; Associated Press on, 136–37; *Atlanta Journal-Constitution* on, 138; *Baptist Press* on, 137–38, 142; vs. the Call to Action, 148–51 (*see also* "Climate Change: An Evangelical Call to Action"); *Christian Index* on, 137–38, 141–44; *Christian Science Monitor* on, 139; Climate Change Interpretation of, 133, 137–41; CNN on, 136, 139; creation of, 133–37, 147, 256n11; vs. the ECI, 140; embattlement perspective on, 150–52; full text of, 229–33; genre of, 139, 145–46; National Public Radio on, 136, 139; news coverage of, generally, 132–33, 136–42, 151, 256n16; *New York Times* on, 132–33, 136, 139; "salt and light" reference in, 145; signatories to, 133–35, 137–38, 141–45, 147, 226–27, 256n12; Stewardship Interpretation of, 133–34, 141–52; textual analysis of, 136–38; *Time* on, 136, 138–39

Southern Baptist Environment and Climate Initiative. *See* SBECI

speaking in tongues (glossolalia), 18–19, 71, 244n78

Spencer, Roy, 175–76

State of the World, 55

Stephens, Randall, 212–13

"Stewardship of the Environment," 127

Stiglitz, D. Joseph, 265n8

Stonestreet, John, 161–62

Strandberg, Todd, 247–48n65

Strauss, Anselm, 102, 226

Sumser, John, 187

Supreme Court, 93–94, 96

Sutton, Matthew A., 244n72

Swanson, Kevin, 166t, 187, 259n14, 264n127

Systematic Theology (Grudem), 186

Taylor, Bron, 242n29

Tea Party, 155–56, 207

Tebow, Tim, 90

Teles, Steve, 191, 202, 266n51

The Blue Zones: Lessons for Living Longer from the People Who've Lived the Longest, 125

theological liberalism, 18, 93, 244n72

There's a New World Coming (Lindsey), 44

think tanks, conservative, 104, 192, 207. *See also* Heartland Institute; Heritage Foundation

Thomashow, Mitchell, 49

thought communities, 115–17

Time, 31, 136, 138–39, 185, 197

Tohoku earthquake and tsunami (Japan, 2011), 250n15

tornadoes, 47

traditionalist evangelicals, 3, 224; characteristics of, 19, 245n80; conservatism of, 217; defining, 16–20, 244n68; vs. modernists, 245n80; Republican, 19; scripture's authority for, 19; size of group, 19; the South as a stronghold of, 20. *See also* Religious Right

Truesdale, Al, 248n68

Truths That Transform, 167t, 174–75, 183, 227

25 Pro-Family Policy Goals for the Nation, 181. *See also* FRC

Tyler, J. Adam, 162–63, 258n7

United States: religiosity and conservative politics in, 216–17, 270n6; role in curbing greenhouse gas emissions, 154; withdrawal from the Paris Agreement, 214

USA Today, 198

Values Voter Summit (2006), 204–5

Van Impe, Jack, 44

Vardiman, Larry, 175–76

Viguerie, Richard, 191–92, 221, 265n9

Vocal Point, 166t, 175

Wagner, Sandra, 263n99

Wald, Kenneth, 35